METHODS IN PLANT BIOTECHNOLOGY

NIPA GENX ELECTRONIC RESOURCES & SOLUTIONS P. LTD.
New Delhi-110 034

About the Author

Kirpa Ram Koundal was born in the village Jamrela, District Kangra, Himachal Pradesh. He did his schooling from Government Senior Secondary School, Bhullana, Kangra, H.P. He obtained his B.Sc. (Agriculture) Hons. degree from Punjab Agricultural University, Ludhiana, Punjab, in '70, M.Sc. (Biochemistry) from Indian Agricultural Research Institute, New Delhi in '72 and Ph.D. (Plant Biochemistry) specializing in the field of plant molecular biology on a national scholarship, Government of India from the University of Alberta, Edmonton, Canada in '77. He joined IARI as an ARS Scientist in '78 and was appointed Professor of Molecular Biology & Biotechnology in '93. He was selected as Director, ICAR-National Institute of Plant Biotechnology, New Delhi in 2000 and as Joint Director (Research), Indian Agricultural Research Institute, New Delhi in '06. He was awardees of ICAR, CSIR and UGC Emeritus Scientist positions at the NIBP, New Delhi after his retirement. His fields of specialization are plant biochemistry, plant molecular biology and biotechnology. Prof. K.R. Koundal has made outstanding contributions in research, teaching and training in the field of plant molecular biology & biotechnology. As a Principal Investigator of ICAR Project on Isolation of Plant Genes of Agronomic Importance, he has isolated and characterized one dozen seed storage protein and insect resistance genes and their promoters from various legumes and indexed in the NCBI GenBank and developed transgenic tobacco, mustard and pigeon pea plants resistant to insect-pests. As a Professor and Principal Investigator of ICAR-Team of Excellence Project for Human Resource Development in Plant Biotechnology, he has contributed significantly in skilled Human Resource Development by teaching and training the basic tools and techniques in the field of molecular biology and biotechnology to the scientists of the various institutions.

K.R. Koundal was awarded INSA-Royal Society (U.K.) fellowship for Post-Doctoral Position at the University of Durham, U.K. in '85, FAO Fellowship as a Visiting Fellow at the University of Virginia, Charlottesville, VA, USA in '91, and World Bank Fellowship as an Adjunct Professor at the University of Washington, Pullman WA, USA in '98. He was awarded Dr. R.D. Asana Endowment Prize for the triennium 1986-89, Best Teacher Award of IARI in '96 and Bharat Rattan Dr.Chidambaram Subramaniam Outstanding Teacher Award of ICAR in '97 at the national level. Prof. K.R.Koundal was member of the Review Committee on Genetic Manipulation (RCGM), Task Force on Agricultural Biotechnology and Task Force of Basic Research Committees of the Department of Biotechnology, Govt. of India. He was nominee of GEAC for monitoring cotton and mustard GM crop field trials. He has been member of QRT and RAC of several ICAR institutes and State Agricultural Universities in the country. Prof. K.R.Koundal has taught three major post-graduate courses in the discipline of Molecular Biology and Biotechnology; Nucleic Acids, Protein Synthesis and Plant Genetic Engineering including practicals and guided 24 IARI Post-graduate (8 M.Sc. and 16 Ph.D.) students in the field of plant molecular biology & biotechnology and also several post graduate students and post-docs from various universities during his 35 years of service at IARI, New Delhi. He has written 18 book chapters and published 109 research papers in the field of plant biochemistry, plant molecular biology and biotechnology in the national and international journals of repute.

METHODS IN PLANT BIOTECHNOLOGY

K.R. Koundal, M.Sc., Ph.D.
Former Professor
Molecular Biology and Biotechnology
&
Former Director, ICAR-National Institute of Plant Biotechnology
Indian Agricultural Research Institute Campus
New Delhi-110 012

NIPA GENX ELECTRONIC RESOURCES & SOLUTIONS P. LTD.
New Delhi-110 034

NIPA GENX ELECTRONIC RESOURCES & SOLUTIONS P. LTD.

101,103, Vikas Surya Plaza, CU Block
L.S.C.Market, Pitam Pura, New Delhi-110 034
Ph : +91 11 27341616, 27341717, 27341718
E-mail:newindiapublishingagency@gmail.com
www: www.nipabooks.com

For customer assistance, please contact
Phone: + 91-11-27 34 17 17 Fax: + 91-11- 27 34 16 16
E-Mail: feedbacks@nipabooks.com

ISBN: 978-93-90591-72-5

Composed and Designed by NIPA.

Dedicated to
My late Parents
and
My wife late Sharda (Baby)

Preface

Biotechnology is a broad discipline in which biological processes, organism cells or cellular components are exploited to develop value materials for the betterment of mankind.Technological advances influence the practice of science in a major way and available methods often dictate experimental approaches. Within the last three decades the developments of many powerful techniques and tools have resulted from the accumulated knowledge of molecular biology. The recombinant DNA is at the centre of the new advances which have revolutionized biology. These techniques offer new experimental approaches to study complexity of multi-cellular organisms and provide additional information about metabolism, development, regulation and sub cellular structure and function that were impossible earlier.The biologists of developing countries using these techniques to keep pace in the field of plant biotechnology in various aspects including nutrition, biotic and abiotic stresses etc in various crops of economic importance. In developed world, these techniques are easy to practice as various kinds of commercial kits are available. However, in developing countries such kits are either very expensive or not at all available in timely manner at many research institutions at the various places. Furthermore, there are many manuals available but they are very comprehensive giving detailed explanations for each aspect. There is need of a concise manual of basic methods in plant biotechnology with standardized protocols for various techniques for the use of scholars and researchers in the developing countries that prefer local resources. They prepare fresh reagents and solutions of their own and do not heavily rely on the heavy priced kits available from outside for their research work.

Based on my personal experience of teaching molecular biology and biotechnology courses to the post graduate students at the IARI for three decades and realization that that there is a lack of teaching material for conducting practicals in plant biotechnology in the various institutions across the country, the text of this book 'Methods in Plant Biotechnology' has been prepared to provide hands on trainings in various techniques. The efforts have been made to include working protocols of all the basic techniques of recombinant DNA technology giving every precution. The students of biotechnology discipline will find it useful in learning the techniques by performing the experiments given in this practical book. The book includes protocols of all the important methods currently in use in the field of plant biotechnology e.g. isolation of plasmid, cloning and transformation of bacteria, isolation, purification and characterization of nucleic acids, radioactive labelling and probing of nucleic acids, isolation of polyribosomes and *in vitro* protein synthesis, construction and screening of gene libraries, M13 clonng and DNA sequencing, PCR, RT-PCR, RFLP and RAPD, gene maping and genetic transformation of field crops. A practical example of actual research problem in the field of plant genetic engineering; Development and characterization of transgenic crop plants with well defined objectives, conduct of various experiments and results explained with figures are also included. The glossary of important terms given will be helpful to understand the subject matter.The appendix including common conversion, DNA data and preparation of common laboratory solutions used in the various protocols is included in the end of the manuscript.The book includes these brief well worked out protocols with relevant detail in simplified language with precautionary notes which will be helpful to all the postgraduate students and researchers working in the field of plant biotechnology at various institutions within the country. The experiments can be conducted with freshly prepared solutions and buffers without the need of buying expensive imported kits. I have made efforts to cite the suggested references of the original contributors of the techniques in each chapter. I thank my students for sharing their knowledge. I also acknowledge with thanks the help of my colleagues Drs. Rekha Kansal, I.M. Santha, T. Mohapatra and K.C. Bansal in standardizations of some of the protocols. I am indebted to Dr. S.L. Mehta, Former Dean, IARI and DDG (Education) ICAR, New Delhi and my Mentor for his continuous guidance and motivation throughout my academic pursuit. I am grateful to my family members for their kind support. I also thank Mr. Satender Singh for typing the text of this book.

It is hoped that this book entitled 'Methods in Plant Biotechnology' will be useful to all the graduate and post-graduate students and scientists, working in the field of plant molecular biology and biotechnology at various institutions in the country. I look forward for future suggestions / comments and would greatly appreciate receiving them from the users of this book.

Kirpa Ram Koundal

Contents

General Guidelines for Safety in the Biotechnology Laboratory

The protocols require working with living organisms which are potentially dangerous and use of reagents, some of which are toxic and explosive. Carelessness can result in serious injury or illness to you as well to others. For your own safety and that of others, strictly observe the following:

1. Wear a laboratory coat and gloves while working in the laboratory.
2. No Eating, No Drinking, No Smoking, No Mouth pipetting is permitted in the laboratory.
3. Regard all micro-organisms potentially pathogenic and treat them accordingly.
4. Clean your working area with disinfectant before starting and ending your work.
5. Report radioactive spill to the In-charge, no matter how small the spill.
6. Surfaces exposed to microbes should be cleaned with sanitizing agent (70% ethanol).
7. Wear safety spectacles and gloves while viewing ethidium bromide stained gel under UV. The ethidium bromide is extremely mutagenic and UV light is damaging to skin and eye.
8. Make sure that centrifuge tubes are balanced across the centrifuge rotor. Do not open the lid before the rotor has stopped. Use common sense and pay attention to what you are doing.
9. Use all glass-ware and other supplies like Eppendorf tubes, tips etc autoclaved and siliconized for genetic manipulation experiments.
10. Discard all the waste of living organisms, agar plate culture etc only after autoclaving.
11. Make sure all containers bottles you use are autoclaved.
12. All the solutions used should be either autoclaved or filter sterile.
13. Use survey meter to detect any possible contamination.
14. Wear radiation safety badges, gloves and protective eye wears while working with ^{32}P radioisotope.

15. Acrylamide is a potent cumulative neurotoxin.To avoid contact with skin and inhalation of dust and vapors, unpolymerised acrylamide should be handed in a fume hood.
16. Pay special attention when using flammable liquids such as ethanol, isopropanol and chloroform and store these in the fume hood.
17. All the radioactive waste should be disposed at the designated place. These can be stored behind shields to decay prior to disposal at official authorized place.
18. Phenol should be redistilled and saturated with diethyl pyrocarbonate (DEPC) treated water before using for all the molecular biology work.

Glossary

Agrobacterium tumefaciens: A species of gram-negative, rod-shaped soil bacteria; often contains copies of a large plasmid, designated Ti. A Ti-containing strain can infect the stems of many plants and form crown gall tumors.

Anneal: To hybridize nucleic acid molecules by means of complementary base pairing between two strands of DNA or between DNA and RNA.

Antibiotic: A substance produced by one organism, that inhibits or kills another organism, usually bacteria. Antibotics typically are active against bacteria and are produced by fungi or streptomyucetes.

Antisense gene: An engineered gene placed in inverted orientation relative to a promoter which when transcribed, produces a transcript complementary to the mRNA transcribed from the normal orientation of the gene.

Antiserum: Blood serum containing specific antibody against an antigen is an antiserum.

Aspetic: Characterized by absence of contaminating fungi, bacteria, viruses, mycoplasms, and other micro-organisms (e.g. in cultures).

Autoradiography: A method of determining the presence and location of radioactively labeled molecules by their effect in creating an image on a photographic emulsion, usually X-ray film, by activating the silver when film is developed. Autoradiography is commonly used to determine whether a radioactive probe molecule has hybridized to denatured DNA or RNA following Southern or northern transfers, respectively or in colony hybridization procedures.

Base: One of the four chemical units on the DNA or RNA molecules, according to their order and pairing, represent the different amino acids. In DNA, the four bases are adenine (A), cytosine (C), guanine (G), and thymine (T). The purine bases are adenine and guanine; the pyrimidine bases are cytosine, thymine, and uracil.

Base pair: Two nucleotides bases on different strands of the nucleic acids molecules that form hydrogen bonds between them. The bases can pair in a single way only; adenine with thymine (in DNA) or uracil (in RNA), and guanine with cytosine.

Biotechnology: Development of products by a biological process requiring engineering technologies, such as fermentation or controlled environments, or utilizing current technologies (such as recombinant DNA techniques) for the modification and improvement of biological systems including plants.

Blot: To transfer DNA, RNA, or proteins, usually from an electrophoretic gel to an immobilizing matrix such as nitrocellulose, or nylon membranes. Also, the autoradiograph obtained from hybridization analysis after the transfer is called blot.

CAT box or CATT box: A conserved sequence found a part of the promoter region 75 bp upstream of the transcription startpoint of the protein-encoding genes of many eukaryotic organisms.

cDNA: Complementary DNA is a a single-stranded DNA complementary to an RNA molecule. cDNA is synthesized *in vitro* from the RNA primer by the enzyme reverse transcriptase. For cloning purposes, double-stranded DNA can be made from the cDNA template by the enzyme DNA polymerase and inserted into an appropriate vector.

cDNA cloning: The use of cDNA is to clone the coding sequence of a gene or sets of genes by starting with mRNA. The cDNA from mature processed transcripts does not contain introns, the presence of which could act to prevent expression of plant or animal cDNAs when attached to a suitable promoter and cloned into a host organism sch as bacteria or yeast.

Chimeric gene: A recombinant gene that contains sequence from more than one sources such as a coding sequence from one source and a promoter region from another.

Chromosome: The structural component of the cells that carries gene; composed of DNA and proteins.

Clone: A population of recombinant DNA molecules all carrying the same inserted sequence or a population of cells derived from a single cell by mitoses or population of plants derived from a single cell by vegetative propogation (e.g. by cuttings, bulbs, tubers, rhizomes, stolons, or meristem culture) of ro use recombinant DNA techniques to insert a particular gene or other DNA sequence into a vectror molecule.

Coding sequene: The portion of a gene which directly specifies amino acid sequence of its protein. Non-coding sequences of genes include control regions, such as **promoters, operators, enhancers, terminators,** and **introns**.

Constitutive: An organism is said to be constitutive for the production of an enzyme or other protin in that protein is always produced by the cells under all physiological conditions; i.e. the gene is always on.

Copy number: The number of plasmid molecules per cell or the number of times a specific gene is present per haploid genome.

Contransformation: A technique in wherein host cells are incubated with two plasmids, one containing a selectabe marker gene and the other containing a gene that cannot be identified by direct selectin. The cellls selected for the presence of the marker gene are competent for DNA uptake and are more likely to have also been transformed with the second plasmid, thereby reducing the number of cells that must be screened by other means (such as proble hybridization for the second gene.

Denaturation: The breakdown of the secondary and tertiary structure of a protein or nucleic acids by physical or chemicals means. More specifically, the conversion of a double-stranded nucleic acid to the single stranded state by destruction of the hydrogen bonds maintaining the double-stranded state.

Electroblotting: The electrophoretic transfer of macromolecules (DNA, RNA or protein) from a gel, they have been separated onto a support matrix such as a nitrocellulose sheet. The procedure is an alternative to the capillary transfer used in techniques such as **Southern and northern blotting**.

Electrophoresis: A technique for separating different types of molecules based on their patterns of movement in an electrical field depending upon their mass and charge.

Electroporatin: Application of an electrical current across a membrane (such as protoplast) inducing temporary pores and permitting uptake of molecules, organelles etc outside environment.

End-labelling: The linking of a radioactive atom such as ^{32}P to the end of a DNA or RNA molecule by a ligation reaction carried out by an enzyme such as T_4 polynucleotide kinase.

Endonuclease: An enzyme that cleaves within a polynucleotide chain

Exon: That part of the gene which is transcribed and represented in the mature mRNA after transcript processing and removal of sequences corresponding to **introns**. The exon sequence information thus is represented by the protein product.

Expression vector: A cloning **vector** containing an effective promoter such that a gene inserted at a specific site will be efficiently transcribed and translated into a protein product.

Extrachromosomal: located elsewhere than nuclear chromosome, such as in an organelle or a plasmid.

Gene: That physical and functional unit of heredity that encodes functional protein or RNA molecule. More specifically, a segment of chromosome, plasmid, or DNA molecule that includes regions preceding and following the coding region, and possibly introns located between exons.

Genetic engineering: The use of *in vitro* techniques to produce DNA, molecules containinig novel combination of genes or other sequences in living cells that make them capable of producing new substances or performing new functions.

b-*glucuronidase (GUS):* A reporter gene from *E.coli* (hydrolase) that catalyses the cleavage of a wide variety of β-glucoronides. One of the most commonly used is 5-bromo-4-chloro-3-indolyl glucoronide (X-GLUC), which shows dark blue colour in transformed cells and tissues.

Homology: The extent to which the two nucleic acid molecules have the same nucleotide sequences or two proteins have the same amino acid sequences. Homology can be determined by direct comparisons of sequence data, or estimated by DNA-DNA or DNA-RNA hybridization.

Hybridization: The pairing of complementary DNA or RNA strands to give stable DNA-DNA or DNA-RNA duplexes. The efficiency of hybridization is a test of sequence homology.

Inducer: A small chemical molecule or a physical agent which increases transcription rate of a specific gene when applied to cells, tissues, organs or organisms.

Intensifying screen: A plastic sheet impregnated with a rare-earth compound (e.g. calcium tungstate). The rare-earth compound absorbs beta radiation and emits lights. When placed on one side of piece of X-ray fils with a radioactive sample on the other side, the intensifying screen will capture some of the beta capture will blacken the X-ray film and enhance the sensitivity of the detection. It is used in **Southern** and **northern** blotting experiments.

Intron: A DNA sequence within a gene that is transcribed, but is removed from the transcript by splicing together exon sequences on either side of it.

In vitro: The process carried out in sterile cultures. outsie the intact organism. Applied to various biological processes (e.g. tissue culture) and measurements (e.g. enzyme reactions).

In vivo: Occuring in a living whole organisms (animal or plant).

Ligase: An enzyme which forms a phosphodiester bond between two adjacent bases separated by a nick in one strand of a double-stranded DNA molecule. Ligases are also used to join two separated blunt end DNAs, and for the joining of RNA.

Library: A collection of cloned DNA fragments. The library may consist of cDNAs or genomic clones.

Nick: A break in DNA molecule, more specifically, the absence of a phosphodiester bond between adjacent nucleotides of the two strands of a DNA duplex.

Nick translation: A technique for radiolabeling DNA molecules to a high specific activity for producing a probe to use in nucleic acid hybridization. Random nicks or gaps are introduced into the probe DNA molecules using a limited DNase digest. The 5' to 3' exonucliease activity of DNA polymerase 1 enlarges the gaps, and at the same time its polymerase activity fills in the gaps with alpha-labelled ^{32}P deoxynucleotide triphosphates to yield newly synthesized DNA stretches that are highly radioactive.

Nitrocellulose: A nitrated derivative of celluloe that is made into membrane filters of definied porosity (e.g. 0.45 μm, 0.22 μm). These filters have a variety of uses in molecular biology, particularly in nyucleic acid hybridization experiments, in which the nucleic acids are transferred from an agarose gel to a nitrocellulose filter.

Northern blot: The transfe of RNA from an electrophoretic gel to a filter so tha it can be hybridized with a nucleic acid probe.

Oligonucleotide: A short-chain nucleic acid molecule consisting of smalle number of nucleotides.

Organelle genome: DNA of chloroplasst or mitochondria. The chloroplast genome ranges in size from 20 to 200 kb, and the mitochondrial genome in higher plants ranges from 200 to >2200 kb. These genomes exist as circular chromosomes lacking structural proteins and may be present in many copies per cell. Small linear or curcular plasmids may also be present in mitochondria.

Plasmid: An autonomous, self-replicating,extrachromosomal circular DNA. The size varies from <1kb to >300 kb and copy number per cell ranges from 1 to >100; have genes for specific nfunctions, such as antibiotic resistance.

Polymerase chain reaction (PCR): An *in vitro* procedure to amplify the number of copies of a specific DNA sequence often found in rare abundance in a complex DNA source. Two single-stranded synthetic oligonucleotides of defined sequence are required to prime synthesis of the source DNA sequence located between its sites of homology to the primers. Primers are annealed to denatured DNA and DNA polymerase from *Thermus aquaticus* is added to initiate synthesis of two complementary DNA strands by extension from the primers. The reaction mixture is alternately heated (to produce single-stranded templates for the next round of synthesis) and then cooled (to allow synthesis to proceed). After the first few cycles' most new strands end at the primer from the previous synthesis. A sequence of a specific length and base sequence can be amplified many orders of magnitude and used for cloning, sequence analysis, and other molecular biology procedures.

Primer: A short oligonucleotide that is paired with one strand of DNA and provides a free 3' hydroxyl group required for initiation of DNA synthesis by DNA polymerase.

Probe: Defined as nucleic acid sequence that has been labeled with a radioactive isotope or fluorescent dye. It is used to identify specific DNA or RNA molecules that have the complementary sequence. Also it is used to perform hybridization to detect a specific gene or transcript.

Promoter: A regulatory regin of DNA near one of the two ends (usually the 5' end) of the **coding sequence** of a gene or operon that binds RNA polymerase and directs the enzyme to the correct transcriptional start site.

Protoplast: A membrane-bound cell formed after the cell wall is removed from a microbial or plant cell by the action of pectinases and cellulases. Protoplasts are used to produce hybrid cells *via* fusion and to facilitate transformation through direct gene transfer.

Recombinant DNA: DNA molecules in which sequences that are not naturally contiguous have been placed next to each other by *in vitro* manipulations. The different sequences placed within a recombinant DNA molecules frequently are obtained from entirely different organisms.

Regulatory gene: Gene which controls protein-synthesizing activity of the other genes.

Restriction endonuclease: An endonuclease that recognizes a specific sequence of bases within a double-stranded DNA. Type-I restriction endonucleases bind to this recognition site but subsequently cut the DNA at approximately random sites, while Type-III resetictio endonucleases cut at about 25 bp from the recognition site. Type-II restriction endonucleases both bind and cut within their recognition or target site. They may make the cuts in the two DNA strands exactly opposite one another and generate **blunt ends** or they may make staggered cuts to generate **sticky ends**. Only Type-II restriction endonucleases are used in recombinant DNA technology. Names for restriction endonucleases start with three italic letters identifying the source bacterium; subsequent letters (not italic) identlfy the strain or serotype; and capital roman numerals at the end of the name indicate enzyme Type-I, II, or III. Not all possible elements appear in every name. For example: *EcoB* is derived from strain B of *Escherichia coli*; *MboI* is Enzyme I from *Moraxella bevis*; *Bam*HI is enzyme I from Strain H or *Bacillus amyloliquefaciens*: *Hind*III is Enxyme III from Serotype d of *Haemphilus infleunzae*.

Restriction fragment: An individual polynucleotide sequence produced by the digestion of DNA with a restriction endonucleases.

Restriction site: The specific nucleotide sequence in DNA recognized by a Type-II restriction endonucleases and within which it makes a double-stranded cut. Restriction sites usually comprise four or six base pairs and have bilateral symmetry. For example:

5’-G A G C T C-3’

3’-C T C G A G-5’

The two strands may be cut directly opposite to one another (to create **blunt ends** or in a staggered manner (giving **sticky ends**), depending on the enzyme involved.

Southern blot: A technique (designated after its inventor, E.M. Southern) which combines the resolving power of agarose gel electrophoresis with the sensitivity of nucleic acid hybridization. DNA fragments separated in an agarose gel are denatured *in situ* and then blotted by capillary action or electric field from the gel onto a nitrocellulose sheet or other binding matrix placed next to the gel. Single-stranded DNA binids to the nitrocellulose and is then available for hybridization with a ^{32}P-labelled or biotinylated, single-stranded DNA or RNA probe. The hybrids are detected by autoradiography in the case of ^{32}P or a colour change (in the case of a biotinylated probe).

Splicinig: Transcript processing by which intron sequences is removed from precurson RNA molecules and adjacent **exon** sequences are rejoined by ligation. In recombinant DNA terminology, splicing is also used to indicate the isertion of a DNA sequence into a **vector** molecular.

T-DNA: The segment of DNA from Ti or Ri plasmid that is transferred from *Agrobacterium* to the genome of its plant host and causes tumor formation or root induction. The T-DNA can be stably inherited by plants regenerated from Ti or Ri plasmid-induced callus tissue. Foreign DNA inserted into the T-DNA can also be stably inherited.

TATA box: A highly conserved, AT-rich, 7-base sequence found about 25 bases prior to the transcription start site or RNA polymerase-II-transcribed eukaryotic genes. The TATA sequence facilitates, but is not essention for transcription.

Ti plasmid: A class of large, conjugative plasmids found in the soil bacterium *Agrobacterium tumefaciens* and is responsible for the **crown gall** disease of broad-leaved (dicotyledonous) plants. The T-DNA region on the Ti plasmid is inserted into the genome of infected host cells or plants, which makes Ti plasmids and T-DNA especially useful for transformation of foreign genes into certain plants.

Transcript: RNA product of a gene. The primary transcripts initially synthesized by RNA polymerases must often be mature, functional mRNA, rRNA or tRNA species.

Transgenic: Of cells, cell cultures, plants, or progeny: having reeived a foreign or modified gene by one of the various methods of transformation as described in this manual.

Transient expression: The short-term detectable expression of product of a gene which has been transferred to protoplasts, cells, or plants. Marker genes on the DNA construct (such as *GUS* or GFP β-*glucuronidase*) that are readily assayed in plant tissues are often used to determine whether DNA has been transferred into the cells.

Transient expression indicates only that the marker gene is transcribed and translated in the cells assayed, and does not imply that the gene or DNA construct has been heritably incorporated in to the plant genome.

Transformation: The introduction of DNA into the cells of an organism by a method other than sexual crossing. Plasmids are commonly transformed into bacterial cells by direct DNA uptake through the cell wall and membrane.

Vector: The plasmid or phage chromosome used to carry cloned DNA in recombinant DNA experiments. A vector generally contains one or more unique restriction sites, is capable of autonomous replication in a defined host, and confers a well-defined, selectable phenotype on the host organism (e.g. drug resistance).

Western blotting: A procedure for transferring of proteins after separation on a polyacrylamide gel to a suitable immobilizing matrix such as a nitrocellulose sheet. The proteins attached to the support matrix can then be probed with a specific antibody to identify a particular protein species.

1

Background - Plant Biotechnology

Plant biotechnology can be defined as the introduction of desirable traits into plants through plant tissue culture and genetic engineering techniques. The recombinant DNA technology is used to modify the expression of genes already present in the plants or to introduce new genes of other species with which the plant cannot be bred conventionally. Foreign genes can be stably introduced and expressed into plant chromosomes by a several techniques. Most plant cells are totipotent and regeneration of a fertile transgenic plant from a single transformed cell is now routine work.The three major techniques have been used to transfer genes of interest in plants. These include *Agrobacterium* mediated approach, protoplast based approach and biolytic or gene gun approach. A gene of interest is isolated and then cloned into a plant transformation vector.The manipulation of the gene is done in *Escherichia coli* prior to its insertion into vectors for transformation into plant tissue. Transfer of the gene to the host plant is accomplished via the plant pathogenic bacterium, *Agrobacterium tumefaciens*. Some plant species are difficult to regenerate from tissue can be transformed by gene gun. Ballistic projection is now often used for delivering DNA to intact plant tissue protoplasts. Tungsten beads coated with DNA are accelerated at high speeds into plant tissue. Plants that are recalcitrant to transformation and regeneration like monocots cereal plants can be transformed by this technique. Dicots are natural hosts for *A. tumefaciens* are easier to transform and regenerate from protoplasts than monocot plants. Many dicot crop plants, including tomato, cotton, tobacco, lettuce, sunflower, canola, sugar beet, alfalfa, flax, soybean, potato, celery, cucumber, carrot, cauliflower, Arabidopsis, petunia and grapes have been transformed. Monocot plants include the economically important grain crops such as rice, corn, sorghum and wheat. They are not susceptible to infection by *A. turnefaciens*. Exogenous DNA to protoplasts of monocots can be delivered with gene gun but it is difficult to regenerate viable plants from the transformed cells. Progress in the regeneration of rye, maize, and rice and ballistic methods of plant transformation during embryogenesis without destroying the embryo offers opprtunity of transforming many monocots.

One reason for transforming crop plants is to confer a trait to improve its agricultural, horticultural, or ornamental value. The second is to facilitate an understanding of a particular biological process. Many traits have been "engineered" into plants by the addition of a single gene. Crops such as tomato, potato, lettuce, canola, cotton, soybean, com, and rice have already been transformed and some of them are commercilized Genetic manipulation of plants is already having a major impact on agricultural practices and productivity. Historical developments of application of plant biotechnology techniques for developing biotech crops or GM crops with some of the important traits are given below:

i) Insect resistance

If functional insecticides can be genetically engineered into plants, it may be feasible to develop crops that are intrinsically tolerant of insect predators. Such plants would not have to be sprayed with costly and hazardous chemical pesticides. In addition, as is the case with many biologically generated insect pathogens, such insecticides should be required to much lower biological concentrations than exogenously applied synthetic pesticides. Moreover, since biological insecticides are usually specific, they are generally not hazardous to the intended consumers of the food.

Two main strategies have been used to confer resistance against insect predators, i.e. first involves transforming plants with a gene for an insecticidal protoxin produced by one of several subspecies of *Bacillus thuringiensis.* The protoxin does not persist in the environment nor is it hazardous to mammals, and thus, it is a safe means of protectin plants. The gene for the toxin has been introduced and selectively expressed in a number of plants species including tomato, tobacco, potato and cotton. The prospects for the successful application of this approach to crop protection appear to be quite good, provided that the over use of *Bacillus thuringiensis* insecticidal toxins does not facilitate the development of resistant insects.

The trypsin inhibitor protein from cowpea has been shown to be effective amongst a wide variety of insects its presence restricts the ability of insects to digest food (i.e. plants) by interfering with hydrolysis of plant proteins by the insect. Since the cowpea trypsin inhibitor is a natural plant protein, it can be expressed in plants without adversely affecting the physiology of the host. Similarly, the gene encoding a plant proteinase inhibitor can be introduced readily by transformation, with the expectation that its expression will not be problematic. One drawback, however, is that relative to the *B. thuringiensis* toxin, high concentrations of cowpea trypsin inhibitor are required for effectiveness. Since most proteinase inhibitors are inactivated by heating, a transformed plant expressing this polypeptide should not be hazardous to humans if the produce is cooked, Furthermore, if expression is limited to tissues that an insect is likely to attack (e.g. roots and leaves) and not present in the tissues intended for the enduser (e.g., fruit.Now, there are various Bt toxins and protease inhibitors isolated from different sources which are used either individually or in combination to develop various crop plants resistant to various pathogens.

ii) Viral resistance

For many transgenic plants, following the expression of a viral coat protein, the ability of the virus to subsequently infect the plant and spread systematically is greatly diminished. Excess coat protein may interfere with virus assembly and/ or inhibit expression of the viral genome. A number of different viral coat protein genes have been introduced into plant genomes, expressed and subsequently found to confer viral tolerance, including tobacco mosaic virus, cucumber mosaic virus, alfalfa mosaic virus, tobacco streak virus, tobacco rattle virus, potato viruses X and Y, and tobacco each virus plants that have been successfully engineered with viral resistance include tobacco, tomato and potato. Although total immunity has not been achieved, reasonably high levels of viral tolerance have been reported. This approach has great potential as the introduction of a coat protein from one virus can often provide tolerance to a spectrum of seemingly unrelated viruses. In addition, transgenic plants carrying the coat protein have been shown to be virally resistant in the field as well as in the laboratory.

Another strategy that has been successful for introducing viral tolerance into plants is to transform the plant genome with an antisense segment of a viral genome. This approach may provide viral tolerance by interfering with either the translation of viral mRNAs or the replication of the viral genome. However, antisense RNA appears to provide less resistance than over expression of viral coat proteins. Moreover, this approach may only provide protection against the specific virus from which the antisense RNA is derived, since strong hybridization may be required for antisense RNA to be effective. Currently, various other modern approaches are being used for developing crops resistant to virus infection.

iii) Herbicide resistance

The development of herbicide resistant plants is of economic importance to agriculture and is an area where genetic engineering is making significant contributions. If one can engineer a crop to be herbicide resistant, the entire field where the crop is grown can be put under an effective programme of weed control. In the early 1980s, the photosystem II herbicides (e.g., atrazine and DCMU) were considered to be very promising targets for conferring herbicide resistance. The site or action for these chemicals were known to be photosystem II reaction center protein; and its gene (psbA, which is chloroplast encoded) from several species had been sequenced. It had also been found that a point mutation in the psbA gene, converting Ser-264 to Gly, yielded plants that were resistant to Atrazine. However, due to the inherent difficulty of transforming chloroplast DNA, this gene did not become a target widely for bioengineering.

With the advent of ballistic methods of transformation, chloroplast DNA can be transformed and atrazine tolerance can be selectively placed into a plant. Unfortunately, atrazine is a particularly noxious herbicide that is both toxic to animals and persistent in the soil, so there may not be a good rationale for pursuing the development of atrazine resistant plants. This notwithstanding, molecular biological methods applied to atrazine and DCMU action have contributed greatly to our understanding both of herbicide action and photosystem I function.

In contrast to atrazine, Glyphosate is an environmentally "friendly" herbicide that is degraded to nontoxic compounds in the soil. Glyphosate inhibits 5-enolpyruvylshi-mate-3-phosphat synthase (EPSPS), an enzyme in the shikimate pathway preventing synthesis of aromatic amino acids. The gene for the enzyme is nuclear encoded and thus can be anipulated readily by the established methods of plant transfornation. An EPSPS gene has been isolated from glyphosate resistant E. coli, given a plant promoter and introduced into plants, yielding glyphosate tolerant plants. Transgenic species carrying resistance to glyphosate have been developed in tobacco, petunia, tomato, potato, cotton and Arabidopsis. Sulfonylurea herbicides inhibit the enzyme acetolactate synthase, thereby blocking the biosynthesis of branched chain amino acids specific mutagenesis of the gene for this enzyme yields variants that are no longer inhibited by sulfonylurea herbicides. In addition, since this gene is nuclear encoded, it is possible to generate transgenic plants containing an acetolactate synthase gene that confers tolerance to these herbicides. Resistance to sulfonylurea herbicides has been transferred to several plant species including tobacco, Brassica and Arabidopsis.

Bromoxynil is a phenolic-type herbicide that acts by inhibiting photosystem. Rather than attempting to modify the chloroplast encoded gene (psbA) specifying the protein site of action of this herbicide (D I protein), the strategy used to confer tolerance has been to introduce an enzyme into plants which can detoxify this compound before it comes in contect with the D I protein. The gene for a bacterial nitrylase, which can inactivate bromozynil, was transferred to tobacco and the resultant transgenic plants were found to be resistant to the D I protein. Differnt approaches are now being used for improving herbicide resistace in the various crop plants.

iv) Stress tolerance

Due to their immobility, plants cannot avoid stresses such as high light, UV irradiation, heat and drought. At the molecular level, stress often induces the production of oxygen radicals. Thus, tolerance of active oxygen species could increase the resistance of plants to many forms of environmental stress. A common form of active oxygen is the superoxide anion, which may be detoxified via conversion to hydrogen peroxide by superoxide dismutase (SOD) and then to water by any of a number of peroxidases and/or catalases. There was an attempt to increase tolerance to active oxygen by transformation of plants with the gene for a CuZn-SOD under the control of a very active promoter (the promoter for a ribulose-1, 5-bisphosphate carboxylase/oxygenase gene). This approach met with only limited success, possibly because superoxide anion is not completely consumed by the CuZn-SOD and could react with hydrogen peroxide to form the highly toxic hydroxyl radical via the Mehlor reaction. However, tolerance to active oxygen has been achieved by transforming tobacco with a Mn-SOD gene under the control of the cauliflower mosaic virus ^{35}S promoter. It is possible that this protein is effective at alleviating oxidative stress because it quantitatively converts superoxide anion to hydrogen peroxide, eliminating potential contributions from the Mehlor reaction. The addition of a functional Mn-SOD gene has also been demonstrated to confer resistance from oxidative stress to both *E.coli* and cyanobacteria. An additional benefit of increasing the Mn-SOD concentration is tolerance to the herbicide methyl viologen. A natural mechanism of plant protection from UV-B stress is the synthesis of UV screening flavonoids in the epidermal cell layer. In leaves, the synthesis of enzymes involved in the flavonoid pathway is under the control of a UV-B responsive promoter.

Many of these genes (e.g. phenylalanine ammonia lyase and chalcone synthase) have been isolated along with their promoters, so it should be feasible to transfer them into plants that are especially sensitive to UV-B irradiation, alongwith stronger promoters, in an effort to provide the plant with greater UV protection. One problem is that flavonoid synthesis involves an extensive biosynthetic pathway, so it may require the transfer of too many genes to be feasible. It is possible, however, that over production of one enzyme which is at a key juncture in the pathway (e.g., chalcone synthase) might be enough to significantly increase flvonoid synthesis. One of the functions of chaperonins appears to be prevention of denaturation of cellular proteins during stress. In particular, the HSP60 and HSP70 classes of proteins may play this role during heat, drought and light stress. Since the genes for many of these proteins have been isolated, it is possible that they could be placed into plants under the control of strong promoters that are induced during a particular form of stress. A greater emphasis is being given using different stretegies to improve this trait which is goverened my multiple genes and major cause of low productivity of several crops.

v) Improved seed storage proteins

Seed storage proteins are generally made up of a limited number of amino acids organized into repetitive structural unit. They are often missing one or more of the amino acids that are essential to human diets, thus limiting the potential food value of a given grain crop. Recently, progress has been made on altering the amino acid composition of seed proteins so that they contain a better spectrum of amino acids (especially those with sulfur or nitrogen containing residues), consequently raising their food value. This can be done in a few ways. First, foreign seed storage protein genes that encode a broad spectrum of amino acids can be transferred into plants with poor quality seed proteins. This has been shown to be a feasible approach in model studies with tobacco; a *phaseolin* gene from a legume was faithfully expressed and the protein product was correctly compartmentalized in transgenic tobacco. Second, it is possible to alter native seed storage protein genes to produce a protein with an improved amino acid composition (i.e. more lysine and/or methionine residues) and then to reintroduce them into the host plant. Unfortunately, these altered proteins are not always properly compartmentalized into protein bodies in the seeds. However, work on the transport and assembly of storage proteins may greatly expand the utility of this approach. A different approach to this problem is to insert the desired amino acid sequences at the hypervariable regions near the carboxy termini of storage proteins. This seems to have less of a perturbing impact on protein assembly into protein bodies.

Another area in which the quality of seed proteins is of importance is in the production of bread wheat. High molecular weight glutenin subunits confer different degrees of elasticity to bread doughs. In particular, the subunit 10 glutenin has longer and more regular structures and as a result, confers better elasticity to the bread. The genes for these proteins could be modified and reintroduced into bread wheats. A significant contribution has been made in all the aspects mentioned above and many other areas are being exploited to improve the protein quality of plant origin.

vi) Altered fruit ripening and flowers

A major problem in the fruit industry is premature ripening and softening, either prior to or during transport to the market place. There are several ripening patterns in tomato. In particular, an antisense polygalacturonase gene introduced into tomato slows the softening process during fruit maturation while allowing ripening to proceed. Another means of slowing postharvest ripening and softening is to interfere with ethylene production. It has been shown that manipulating the gene for enzymes in the flavonoid biosynthesis pathway can result in novel colouring patterns in flower petals. Thus, understanding the regulation of this pathway may ultimately provide the means to specifically design unique and preferred flower patterns. As is the case for fruit, preservation of cut flowers during shipping is of great importance. Recently, there has been progress in the understanding of some of the mechanisms of flower senescence at the molecular level. It is clear that oxygen radicals play a key role in causing cut flowers to deteriorate. Similar to stress tolerance, if one could increase the endogenous oxygen scavenging capacity in cut flowers by over expression of a Mn-SOD gene, under the control of a flowering specific promoters, their shelf-life might increase. This field is becoming very important as the demand for the cut flowers has increased tremendousliy these days.

vii) Development and commercialization of GM crops

Plant Biotechnology has thus made significant contribution in the past twenty years in the development and commercialization of genetically engineered or GM crops with multiple benefits Currentlly, a variety of traits have been introduced in genetically engineered crops which include: Herbicide resistance, Pest resistance, Viral resistance, Slow-ripening of fruits and vegetables, Fungal and bacterial resistance, Abiotic stress tolerance (Drought, salinity, temperature, flooding etc), Quality improvement (Starch, protein and oil), Value addition (vitamins, micro and macro elements), Pharmaceutical and therapeutic proteins, Edible vaccines etc. Transgenic crops expressing foreign genes, which confer added protection against insect pests and viruses, could make a significant contribution to sustainable agricultural. Several countries after adopting cultivation of GM crops with single gene of agronomic importance are now developing plants that can provide multiple benefits to farmers. The crops having multiple genes are either developed through introducing a new gene in already transgenic varieties or through conventional breeding of two single trait lines conferring durable characteristics. Multiple genes in transgenic crops can be useful in developing durable disease resistance, broad insect control spectrum and could lead to potential cost saving advantage to the farmers. Development of transgenic to produce high value pharmaceuticals and edible vaccine is also under progress which requires much more research and development work before commercially viable products will be available. In addition, molecular aided selection (MAS) is now routinely used to enhance the speed and precision of plant breeding.

The first transgenic crop in the world, the slow ripening 'Flavr Savr' tomato, was released for commercial cultivation in 1995 and since then the global acreage under transgenic crops has increased steadily. A total biotech crop area of

172.7million hectares in 2019 was grown and most adopted biotech crops by the 29 countries were soybeans, maize, cotton, and canola. Soybean was the leading biotech crop with 91.9 million hectares that occupied 48% of the global biotech crop area, with a 4% reduction from 2018. This is followed by maize (60.9 million hectares), cotton (25.7 million hectares), and canola (10.1 million hectares). The USA led the biotech crop planting in 2019 at 71.5 million hectares, followed by Brazil (52.8 million hectares), Argentina (24 million hectares), Canada (12.5 million hectares), and India (11.9 million hectares) for a total of 172.7 million hectares, representing 91% of the global area. The stacked products, currently deployed in the US, Canada, Australia, Mexico, South Africa and the Philippines, are an important and growing future trend, which meets the multiple yield constraints of farmers.

The research and development activities in transgenic crops in India have led to the generation of a number of genetically modified (GM) plants which are at different level of development harboring genes of agronomic as well as quality traits. Three Bt-cotton hybrids BT-MECH-12, BT-MECH-162 and BT-MECH-184 developed by Mahyco-Monsanto engineered for tolerance to insect damage were the first transgenic crops approved for commercial cultivation in India in the year 2002 which were cultivated in about 50,000 hectares. Now, the Bt-cotton adoption rate in India has almost stabilized in the past five years at more or less 95%. The 94% adoption rate in 2019 by the farmers who planted 11.9 million hectares of biotech cotton reflects the continuing confidence of the farmers. It is noteworthy that India, the largest Cotton growing country in the world, registered the highest proportion increase in Bt-Cotton area and Bt cotton has been a major contributor to cotton export during this period.

viii) Emerging issues related to GM crops

With the development of GM crops/food, the issues of human safety, environmental safety and their risk assessment have become one of the great concerns. The bio-safety issues require to be addressed on priority basis. It is of crucial importance that transparent and science based approaches are used to conduct safety assessments on biotechnology products before their commercialization. It is important that correct information regarding the new technology should be communicated properly and effectively to the general public and stakeholders. It is imperative to build public confidence through proper communication system so that the public at large is well informed about the benefits and risk if any, of the GM foods. We will not succeed in developing and deploying GMOs if the public at large is not part of the process as they must understand that this is not just a technical imperative to feed the poor and the hungry but has become an ethical and moral imperative. The issues associated with biotech crops need to be addressed and integrated into economic growth and environmental strategies through acceptance by various stakeholders. Following are the major issues, which assume great importance for commercialization of transgenic crops.

One of the important concerns related to transgenic food is that an allergen (a protein that causes an allergic reaction) could be introduced into a transgenic food product. A variety of questions and tools must be considered to remove this apprehension and science based risk assessment of GM food is the only way to address public concerns. Concerns have also been raised that marker gene (gene that is used as marker to select the GM cells from a mass of new cells to finally produce GM plant) that could move from GM crops to the microorganisms that normally reside in a person's gut and lead to an increased the antibiotic resistance. Efforts should also be made to remove the marker genes already present in the transgenic. Alternative strategies are now being followed for development of transgenic free from marker genes. It is advisable not to have any antibiotic resistance gene present in the GM crops.

A major environmental concern associated with GM crops is their potential to create new resistant weeds, out crossing with wild relatives or persisting in the wild themselves. That an herbicide tolerant gene may escape through pollen into nearby farms and fields, to other GM cultivars or to a wild and weedy relative is a possibility. The genes may escape through pollen into the nearby farm to another non GM cultivar or wild weedy relatives. It is therefore important to evaluate individual GM crops on a case by case basis and the environmental effect of transgenic crops should be assessed and compared with non-transgenic crops grown with conventional practices. The potential transfer of a transgene to local flora and it's any possible subsequent impact on specific plant species is thus an issue that needs to keep in view before commercial release of specific transgenic crop. Another environmental concern with Bt crops is that it will lead to the development of insect resistance to Bt. Therefore, scientists have to develop strategies like pyramiding of insecticidal genes with divergent mode of action to overcome this problem. Additional resistance management practice is also need to be developed to provide durable resistance.

Many consumers feel and insist on their right to know what they are eating and their right to choose and therefore GM food often touches on the issue of labeling GM products. In India also labeling of GM food is mandatory as a measure of consumer safety and consumer information and right of choice. A draft notification issued by MoH & FW, Govt. of India, provided that GM materials shall be compulsorily labeled without any exception. While it is easy to detect GM ingredients in product when the GM gradient is the main ingredient (popcorn), it would not be so easy to detect them in processed products like oils, sugar and starch which no longer contain any novel DNA or proteins. Nevertheless, it is desired to have labeling to the maximum extent possible so that public is informed to know and choose. In case of imported GM food, the importer is also required to submit the reuifred documents at the time of import.In addition, risk assessment preformed should be standardized for plants new to an environment. Regulatory authority should explain the tangible benefits, be honest about the risks, and explain the risk management plan. Consumer response depends on perception about risks and benefits of biotech crops. It is possible that food safety evaluation of GM plant/food may reveal some unavoidable effect. Therefore, in order to maximize the trust, it is essential that relevant and reliable information about the genetically modified food is communicated to the consumers and stakeholders. A participatory approach is immensely required if the biotechnological products are to be accepted by the farmers and consumers. There is need to ensure that the guidelines on GMOs are well advertised to increase public awareness on the subject and to ensure that technology is well received. Ultimately credible evidence from scientists or regulatory institutions will not influence popular public opinion unless there is public confidence in the institution certifying the GM food.

ix) Future prospects

The above given examples of application of plant biotechnology are just a few developments in transgenic technology which provide a broad base background for crop improvement. The production and productivity of several crops continue to be adversely affected due to biotic (pathogens and diseases) and aboitic (salinity, drought, temperature and flooding) stresses and climate change in our country. It is important that we should enhance farm productivity per unit land, water and capital without harming the ecosystem in the present scenario. The plant biotechnology has provided enormous potential in regards to revolutionize agriculture. In recent year, development of transgenic crops expressing a variety of novel traits such as insect resistance, disease resistance, herbicide tolerance, hybrid production, improved oil quality etc have led to large scale cultivation of GM (genetically modified) crops in different countries. In India development and commercialization Bt cotton has been a successful example. Several biotech tools and techniques, including tissue culture, diagnostics, genomics, molecular marker-assisted selection (MAS), and genetic transformation are being used collectively for speeding the plant breeding and help mitigate the effects of climate change. It is important to isolate the genes and promoters of agronomic importance which can be expressed in different crop plants for strengthening the development of transgenic crop resistant to biotic and abiotic stresses and post harvest losses. As the technology improving and our knowledge of gene technology further increasing, it will be easier to transform plants with more than one gene in order to engineer more complex traits of economic importance. The understanding of manipulation of the key steps of the pathways provides the ability to produce a wide range of specific bioactive compounds in plants. Many more plants will be regenerated from tissue culture and a much better understanding of biochemistry and molecular biology of whole range of plant specific processes will be helpful for genetic transformation of recalcitrant plants. Newer technologies need to be developed and deployed for enhancing and sustaining agricultural productivity. There is a need to optimize the use of biotechnology in conjunction with conventional technologies to achieve higher productivity with fewer resources. Therefore, genetic modification of crop plants assumes greater importance which demands development and adoption of newer technology for genetic improvement of crops for various traits for increasing crop productivity.

It is essential that our country adopt all technological advances including genetic transformation based on social and scientific considerations. Several public and private sector institutions are involved in research and development of GM crops in India and models need to be worked out for public and private partnership for successful deployment of technology for farm level applications and commercialization. The concerns associated with the use of GMOs may differ depending on the particular gene-organisms combination and therefore case by case approach is required for assessment of safety concern by each country. The future course of Indian agriculture will depend on the farmers' acceptance or reservations for the adoption of biotechnological applications for crop improvement. There is a great challenge for evolving strategies in a participatory mode to create farmers' awareness for popularization and commercialization on GM crops in our country. This will require making appropriately aware of the scientific merits, apprehensions and precautions for the use of biotech crops. The Government must ensure continued safe and effective testing and introduction of transgenic crops and implement regulatory programs that inspire public confidence. It is important to remember that biotechnology tools complement and extend the traditional methods of plant breeding used to enhance agricultural productivity. We must ensure that farmers get benefit from the vital contribution that plant breeding offers using both conventional and plant biotechnological tools.

Suggested readings

Beachy, R.N. (1990). Plant transformation to confer resistance against virus infection, in gene in plant development, vol 2, Gustafson, J.P., Ed., Plenum Press, New York, 305.

Boyer, J.S. (1982). Plant productivity and environment, Science 218: 443-448.

Chris Bowler, Luit Slooten, Saskia Vandenbranden, Riet De Rycke, Johan Botterman, Chris Sybesma, Marc Van Montagu and Dirk Inze (1991). Manganese superoxide dismutase can reduce cellular damage mediated by oxygen radicals in transgenic plants. EMBO journal 10: 1727-1732.33.

Christou P., Ford T.L. and Kofron M. (1991). Production of transgenic rice [*Oyza sativa* L) from argonomically important indica and japonica varieties by electric discharge particle acceleration of exogenous DNA in to immature zygotic embryos. Bio/technology 9: 957.

Clive James (2019). Global status of commercialized biotech/GM crops-2019. ISAAA. Brief 55.

Corner, T. (1993). Food safety issues relating to genetic engineering of crop plants. Agriculture Science 6: 36-41.

DBT (1998). Revised guidelines for research in transgenic plants and guidelines for toxicity and allergenicity evaluation of transgenic seeds plants and plant parts. DBT, Govt. of India, New Delhi.

Hilder, V.A. and Boulter D. (1999). Genetic engineering of crop plants for insect resistance - A critical review. Crop Prot. 18: 177-191.

Hoffman L.M., Donldson D.D. and Herman E.M. (1988). A modified storage protein synthesized, processed and degraded in the seed of transgenic plants. Plant Mol. Biol. 11: 717.

Jhansi Rani S. & Usha R. (2013). Transgenic plants: Types, benefits, public concerns and future. Journal of Pharmacy Research, 6: 879–883.

Juma C. (2011). Preventing hunger: biotechnology is the key. Noture 479: 471-472.

Kahl, G. and Writer P. (1995). Plant genetic engineering for crop improvement. World Journal of Microbiology & Biotechnology 11: 449-460.

Klumper, W. and quaim, M. (2014). A meta-analysis of the impacts of genetically modified crops. Plos one 9(11): e1111629.

Koundal, K.R. and Rajendran R. (2003). Plant Insecticidal Proteins and their Potential for Developing Transgenic crops Resistant to Insect Pests. Indian Journal of Biotechnology 2:110-120

Koundal, K.R. and Sharma T.R. (2006). Biosafety concerns associated with GMOs. In Biosafety: issues and challenges. P 83-89 (Desh Deepak Verma & Manoranjan Hota Eds), Publication, Ministry of Environment and Forests, Govt of India, New Delhi.

Koundal, K.R. Santha I.M., Mehta S.L. and Malik V.S. (1989). Plant Genetic Engineering-Prospects of Improving Crop Productivity. In Recent Advances in Plant Biochemistry, p. 437-461 (S.L. Mehta, M.L. Lodha and P.V.Sane Eds) ICAR Publications, New Delhi.

Lawrence, P.K. and Koundal K.R. (2001). Agrobacterium tumefaciens-mediated transformation of pigeonpea/Cajanus cajan) and analysis of regenerated plants. Current Science 80: 1428-1432.

lngo Putrikus (2012). "Golden Rice" A GMQ product for public good and the consequences of GM regulation. 1. Plant Biochem. Biotech 21(Suppl 1): S68-S75.

Mudiyappanayar Jagdish and Koundal K.R. (2020). Constitutive expression of protease inhibitor gene isolated from Black gram (*vigna mungo* L.) confer resistance to *sapodoptera litura* in transgenic tobacco. Indian Journal of Biotechnology 19(2): 94-101.

Qaim, M. (2010). Benefits of genetically modified crops for the poor: household income, nutrition and health. New Biotechnology 27: 552-557.

Qaim, M. and Zilberman D. (2003). Yield effects of genetically modified crops in developing countries. Science 299: 900-902.

Ryan C.A. (1990). Protease inhibitors in plants: genes for improving defenses against insects and pathogens. Annual Review of Phytopathology, 28: 425-449.

Saul M.W. Shillito R.D. and Nagrtiu L. (1988). Direct DNA transfer to protoplasts with and without electroporation, in Plant Molecular Biology Manual A1 Kluwer Academic Publishers.

Schuch, W., Bird C.R., Ray J., Smith C.J.S., Watson C.F., Mores P.C. Gra J.E., Arnold C., Seymour G.B., Tucker G.A. and Grierson D. (1989). Control and manipulation of gene expression during tomato fruit ripening. Plant Mol. Biol. 13: 303.

Sharma, H.C., Crouch, J.H., Sharma, K.K., Seetharam, N. and Hash, C.T. (2002). Applicaton of biotechnology for crop improvement. Prospectus and constraints. Plant Science 163: 381-395.

Sharma, M., Charak, K.S. and Ramanaiah, T.V. (2003). Agricultural Biotechnology Research in India: Status and policies, Current Science, 84(3): 297-302.

Sushma Rani, Sharma Vinay, and Koundal K.R. (2017). Efficient genetic transformation of Brassica juncea with lectin gene using cotyledon explants. International Journal of Advanced Biotechnology Research 7: 1-12.

Sushma Rani, Sharma Vinay, Bhattacharya R.C. and Koundal K.R. (2017). Fusion gene construct preparation with lectin and protease inhibitor genes against aphids and genetic transformation of Brassica juncea using cotyledon explants. Acta Physiol. Plant. 39: 115.

Tome D.T., Weissinger A.K., Rosss M., Higgins R., Drummond B.J., Schaaf S., Melone-Schoneberg J. Staebell M. Flynn P., Anderson J. and Howard J. (1990). Trasgenic tobacco plants and their progeny derived by micrprojectile bombardment of tobacco leaves. Plant Mol. Bio. 14: 261.

Vain, P. (2006). Thirty years of plant transformation technology developments. Plant Biotechnology 15(2): 221.

Wu R., Kremmerer E. and McElrory D. (1990). Transformation and regeneration of important crop plants, rice as the model system for monocots, in gene manipulation in plant improvement. Vol 2, 251,Gustafason J.P. Ed., Plenum Press, New York.

Yoder, J.I. and Goldsborough, A.P. (1994). Transformation systems for generating marker free transgenic plants. Bio/ technology 12: 263-267.

2

Isolation, Restriction and Agarose Gel Electrophoresis of *E.coli* Plasmid DNA

Plasmid is a double stranded, self replicating, circular extra chromosomal DNA of bacterium. It is used in recombinant DNA experiments to clone genes from other organisms and make large quantities of their DNA. Plasmid can be transferred between same species or between different species. The sizes of plasmids range from 1-1000 kbps. Plasmids are part of mobilomes (total of all mobile genetic elements in a genome) like transposons or prophages and are associated with conjugation. Even the largest plasmids are considerably smaller than the chromosomal DNA of the bacterium, which can contain several million base pairs. After new plasmids are constructed they may be isolated and characterized with respect to their size and restriction enzyme patter by gel electrophoresis.

There are various methods for preparing plasmid DNA; however the most convenient method of preparing plasmid DNA is the alkali method of Birnboim and Doly (1979). In this method the cells are lysed by SDS at high pH and then neutralized. The plasmid DNA is renatured but the chromosomal DNA remains denatured and precipitate as a protein –DNA-SDS complex. The procedure can be scaled up. However, for sample more than 10 ml, it is best to deproteinise the plasmid with pheny/chloroform prior the ethanol precipitation. For high quality Plasmid DNA preparation the plasmid can be further purified by CSCl equilibrium density gradient centrifugation large and small plasmid DNAs have been extracted by this method.

i) Plasmid DNA miniprep

Reagents

Solution-I

20% glucose	2.25 ml
0.5M EDTA (pH 8.0)	1.00 ml
lM Tris (pH 8.0)	1.25 ml

St H_2O	45.50 ml

Store at 4°C, add 2 mg/ml lysozyme immediately before use.

Solution-II

10 NaOH	0.4 ml
20%SDS	1.0 ml
Sterile H_2O	18.6 ml

Solution-III

3N Sodium acetate, pH 4.8

Make up by dissolving 3 moles NaoAc in water. Add glacial acetic acid to pH 4.8. Bring to 1 litre with water. Store tightly capped.

Solution-IV

RNase A 10 mg/ml in 10 mM Tris pH 7.9, 15 mM NaCI. Heat to 100⁰C for 15 min and store in small aliquotes at -20C.

LB Broth (1 litre)

Bacto tryptone	10 g
Bacto-yeast extract	5 g
NaCI	10 g

Adjust to pH 7.5 with sodium hydroxide; make up volume to 1 litre and autoclave.

Protocol

1. Inoculate 2 ml of LB broth containing appropriate antibiotic with a single colony of an *E.coli* strain carrying a plasmid and grow overnight at 37°C.
2. Transfer 1.5 ml of the overnight grown culture into a sterile Eppendorf tube and spin for 20 sec in a microfuge at full speed. Pour off the supernatant.
3. Suspend the pellet in 100 µl of solution I containing 2 mg/ml lysozyme. Vortex well and keep on ice for 5 min.
4. Add 200 µl Solution II. Mix gently till solution clears by inverting tubes. Keep on ice for 5 min. The contents of the tube will turn colourless and viscous.
5. Add 150 µl of solution III and mix gently by inverting tube several times and keep on ice for 5-10 min.

6. Spin the tube in a microfuge for 5 min, transfer supernatant to a fresh 1.5 ml microcentrifuge tube and discard the petllet.
7. Add two vol of 100% enthanol to the supernatant, keep at -20⁰C for 15 min, spin and discard the supernatant.
8. Dissolve the pellet in 20 µl TE buffer

Note: Treat with solution IV @ 1 µl RNase A (1 mg/ml) per µg DNA during restriction digestion to remove RNA, if required.

ii) Large scale isolation of plasmid DNA from *E. coli*

Media and Solutions

LB medium, Per litre

Becto tryptone 10g; Bacto yeast extract 5 g; Adjust pH to 7.5 with NaOH. Autoclave.

Solution-I

50mM Glucose; 25 mM Tris -Cl (pH 8.0); 10 mM EDTA; Lysozyme solution 10 mg /1 ml in Solution I

Solution-II

0.2 N NaOH; 1% SDS

Solution-III

3M Sodium acetate (pH 4.8)

TE Buffer

10mM Tris-Cl pH 8.0; 1 mM EDTA

Protocol

1. Grow *E.coli*. Strain carrying plasmid overnight.
2. Inoculate 50 µl of the overnight culture into 500 ml LB containing selective antibiotic and grow overnight at 37°C.
3. Harvest the bacterial cells by centrifugation at 4000 g for 10 min at 4°C.
4. Resuspend the bacterial cells in 10 ml of cold solution I.
5. Add 20 ml of solution II. Mix by inverting the tube. Keep at 4°C for 15 min.
6. Add 15 ml of 3 M sodium acetate. Mix by inverting the tube. Keep at -20°C for 90 min.

7. Filter through Mira cloth and centrifuge the filtrate at 5,000 g for 15 min.
8. Add 180 ml isopropanol. Mix well and centrifuge at 5,000 g for 10 min.
9. Redissolve the pellet in 6 ml TE.
10. Add 800 μl of EtBr and 1.1 g of CsCl per ml solution. Centrifuge.
11. Completely fill Becknan's quick seal tube and heat seal.
12. Load the tube on Ti-65 type rotor.
13. Centrifuge at 60,000 rpm at 20°C overnight.
14. After the run carefully remove the tube, and examine in the dark under UV light. At this stage two bands should be visible, an upper sharp band of chromosomal DNA and a lower diffused band of plasmid DNA.
15. Puncture the top of the tube and using a hypodermic syringe and needle, inserted through the side of the tube below the lower band, draw off the DNA of lower band and transfer to a sterile tube.
16. Remove the EtBr completely by partitioning aqueous phase several times with n-butanol saturated with TE.
17. Transfer the plasmid solution to a dialysis tubing, dialyse the solution at 4°C overnight against 2 litres of TE on a magnetic stirrer with 3-4 changes of the buffer.
18. Transfer the plasmid DNA solution to a suitable vial. Add 0.1 vol of 3 M sodium acetate solution and mix.
19. Add 3 volumes of ethanol mix well and keep at -20°C for 2 hrs.
20. Centrifuge at 10,000 g for 10 min.
21. Decant the supernatant, wash the ppt with 70% ethanol.
22. Dry the DNA pellet in a vacuum disiccator.
23. Redissolve the ppt in 200 μl of the buffer.

iii) Restriction of plasmid DNA

Enymes which cut DNA at specific sites are called restricting endonucleases. Although, there are three types (I, II, III) of restriction endonucleases, type II enzymes which recognize specific 4-8 bp long sequences and cut DNA of any source within this sequence are mostly used. These produce cohesive terminii on restriction. The assay conditions vary from enzyme to enzyme.

Plasmd DNA prepared by miniprep method is not pure and a large amount of contaminating RNA is present. To overcome this, RNAse is added to the restriction mixture. In addition, Spermidine can be added to enhance the action of the restriction enzymes.

Requirements

1. Siliconised Eppendorf tubes.
2. 10 x low, medium and high salt buffers.
3. Restriction enzymes.
4. Sterile distilled water.
5. DNA dye i.e. (3.125 ml glyercol; 2.0 ml 50 mM Tris-Cl, pH 7.5, 0.5 ml of 0.2M EDTA, (pH 5.0), 4.375 ml of H_2O; 20 mg agarose and 10 mg of bromophenol blue). Autoclave and squeeze through a syringe to form the beads).
6. In addition to normal restriction requirements the following is optional :Supermidine tri-hydrochloride solution. Make up a stok of 25 mg in 1 ml of sterile water. Store at -70°C until use.
7. RNAse solution: Prepare a stock of 10 mg/ml in sterile water and boil for 10 min. Dispense 100 µl volume in the Eppendorf tubes and store at -70°C until use.

Protocol

1. Add the following ingredients to a siliconised Eppendorf tube (for each restriction) in the following order.

a) DNA (1 µg/ µl) to the bottom of the tube	1 µl
b) Sterile water to the side of the tube	14 µl
c) 5 x Salt buffer to the side of the tube	4 µl
d) RE (15 U/ µl) 1 µl to the side of the tube	1 µl

2. Confirm that addition has been made by ticking the above items as you make the additions. Mix the ingredients by flicking the Eppendorf tube.
3. Briefly centrifuge to bring the liquid to the bottom of the tube.
4. Incubate at 37°C for 2-3 hrs to complete the restriction,. Unrestricted samples to be used as controls are kept on ice. They contain the appropriate amount of DNA made upto 20 µl using TE buffer. Whilest waiting for the restrictions, prepare the agarose gel, as detailed in the next protocol.
5. At the end of the incubation, place the Eppendorfs tubes in water at 70°C for 5 min. This helps to denature the enzymes. Most enzymes are destroyed by this process, the few remaiing viable. For sample analysis, it is not important that all enzymes are fully denatured. Place the samples directly into ice at the end of the 70°C incubation. Leave in ice until required for gel electrophoresis.

Note: Estimate the relative quantities of DNA, water, 5x salt buffer, and enzyme(s) required for each restriction. The enzyme solutions are in 50% glycerol, whilest the maximum level of glycerol permitted in the restriction samples is 5% (to avoid random restriction). Thus, for 20 µl final volume the permitted total volume, for all enzymes is 2 ul (unless the total volume is increased). Excessive restriction enzymes in sample may negatively affect the result.

iv) Agarose gel electrophoresis

Agarose gel electrophoresis is conducted in a horizontal configuration. This configuration provides better support of low gel concentrations, is easier to mainpulate, and there is less distortion (collapse) during electrophoresis. Gels will be submerged by about 1 mm in the electrophoresis buffer. The resistance of the gel is not much greater than the buffer, and thus a large fraction of the current is carried in the agarose.

There is considerable variability with different sources of agarose with respect to the solidity, resolution and mobility of DNA bands, and interferring contaminants. Pure agarose consistantly gives excellent results. The agarose is dissolved in electrophoresis buffer by boiling in a microwave. The solution is cooled to about 50^{0}C (a temperature at which one can hold the fllask, but it is uncomfortable) before pouring the gel. This cooling process may be speeded by melting the agarose at twice the desired concentration. An equal volume of buffer at room temperature is added to the molted solution to bring the agarose concentration to 1x. Alternatively, the agarose may be cooled in a water bath. Pouring the gel when the agarose is too hot will ruin the gel apparatus.

Agarose gel electrophoresis is used to separate DNA fragments by size. The negatively charged DNA molecules migrate through the pores of gel matrix in an electric field from the negatively charged cathode to the positively charged anode, with smaller molecules migrating faster than the bigger ones.

The mobility of DNA is inversely proportional to the log of the molecular length. DNA standards are routinely included. The rate at which the DNA travels through the gel matrix is determined by the voltage gradient (V/cm). Voltage gradients normally used are between 0.5 and 5 V/cm. High resolution, especially of high molecular weight DNA is achieved at low voltage gradients. Higher voltage gradients are used on small DNA (~2 kb) to increase the mobility, thus reducing the total amount of band diffusion. Most commonly use 0.8%agarose at about 1-3 V/cm in TAE buffer.

DNA is visualized by staining with ethidium bromide and viewing under a 300 nmUV light source. Ethidium bromides intercalated into the DNA will fluorece at about 590 nm (orange). The gel may be stained by soaking for about 15 min in ethidium bromide at a concentration of about 0.5 mg/ml.Alternatively, the gel may be prestained by incorporating ethidium bromide at 0.5 μg/ml in the gel and running buffer. Pre-staining results in altered mobility of the DNA. This is especially pronounced for supercoiled and open-circle molecules.

Reagents

DNA dye (composition as given in previous protocol).

1X TBE buffer (pH 8.3)

10X TBE buffer	100 ml
Ethidium bromide (10 mg/ml)	100 μl
Water to make up	1000 ml

Protocol

1. Make 0.7% agarose by adding 1.05g agarose to the flask containing 135 ml water and boil to dissolve agar and cool to 55°C, before pouring gel.
2. Add 15 ml of 10X TBE buffer and 15μl of ethidium bromide solution. Swirl to mix but avoid bubbles.
3. Pour agarose gel solution into gel former mold to make gel.
4. Place well forming comb near one edge of the gel. Clip comb to bridge above gel and make sure fingers of comb are slightly above the plate, but not touching it.
5 Let it harden until it becomes milky and opaque (approximately 30 min.)
6. Add agarose beads DNA dye to the restricted samples, and also to any unrestricted samples under test. Use 2 μl dye per 20 μl of sample or equivalent. Gently mix. Level off the gel tank using a spirit level. Remove the comb from the gel. Place the gel in the tank. Pour in the buffer solution to 2 mm above gel. Remove bubbles from walls with a pipetman.
7. Apply the DNA samples carefully into the individual wells. Change the tip between each operation. Replace the lid. Connect the apparatus to the power supply such that the positive (anode) terminal is connected to the side without the wells. DNA being negatively charged moves to anode. Turn power on and increase the voltage to 30 volts for an overnight run and 130 volts for a run of 2-3 hrs. Check the current. The display is in mAmps. Accept this to be numerically slightly higher than the voltage i.e. 150 mA at 130 volts. High Currents are a Sign of Too much Electrolyte, or a Badly Levelled Tank.

8. At the end of the run, the marker band should be 80% down the gel. Switch off the power supply
9. Photograph ethidiumbands in gel using a UV-transilluminator. Gel is ready for Southern blotting or isolation of fragments.

Note: 0.7% agarose gel is good for DNA pieces that range in size from 1 to 25 kb. For smaller pieces of DNA (200-2000 bp) use 1.5% agarose, for larger DNA size (20-700 kb) use 0.5% gel.

v) Isolation of DNA fom low melting temperature agarose gel

Protocol

1. Prepare an agarose gel (0.7%) using low melting point (LMP) agarose.
2. To 0.42 g LMP agarose, add 50 ml water Head until dissolved. Then add 6ml 10 x TBE uffer. Make upto 60 ml final volume with water. Add 12 µl of ethidium bromide 10 mg/ml in sterile water.
3. Place the gel in the tank whilst comb and the gel surround are still in position. Cover the gel with buffer prior to removing comb and the surround.rose
4 Prior to loading the samples, the gel should be pre-run for a period of 10 min. During this time the voltage should be slowly increased until the required setting is obtained (This prevents splitting of the gel). Load & AA sample, run the gel at upto 7.5 v/cm until the dye migrates 2/3 distance.
5. To remove DNA view the agarose gel under UV light. Care is needed when removing the gel from the plate as the LMP agarose does not have the strength of the ordinary agarose. Remove the required fragments with a sterile scalpel.
6. Melt agarose then by heating it to 65°C and add 2 volumes of 50 mM Tris-HCl, 0.5 mM EDTA, pH 8.0, mix and place at 37°C for a few min.
7. Add an equal volume of phenol, vortex and then centrifuge for 5 min. Remove the upper aqueous layer and re-extract with phenol once. Then re-extract with phenol once. Then re-extract with a solution of chloroform: isoamyl alcohol (24:1).
8. Add 0.1 volume at 3M sodium acetate and then 3 volumes of ethanol. Leave it at -20°C for 1 hr and collect the precipitate by centrifugation (10 min).
9. Wash the precipitate with 80% ethanol, dry under vacuum for 10 min. and re-suspend in sterile water or TE buffer.

vi) Isolation of DNA from agarose gel by electroelution

Protocol

1. Restrict the DNA with the appropriate restriction enzyme(s) according to standard protocols. Electrophorese the restricted DNA in an appropriate agarose gel.
2. View the gel in UV light and cut out the desired band with a sterile scalpel or razor blade. Trim off the excess gel.
3. Take a 4 cm length of narrow dialysis tubing (0.8 cm diam.) washed thoroughly and treated by boiling in 10 mM EDTA for 10-15 min. Secure one end of the tubing by folding it and clipping shut with a Mediclip (alternatively tie two or three knots in it).
4. Insert the gel slice into the tubing and add no more than 0.5 ml of electropficresis buffer (sterile if pos sible) per slice of gel. Exclude all air from the dialysis tubing and close off the other end with another Mediclip or knots.
5. Add 50 ml of electrophoresis buffer into a mini gel tank and submerge the gel in the tubing at right angles to the direction of the current. Also add 5 µl ethidium bromide.
6. Electrophorese at 60 mA for about 20 min or until all the DNA has left the gel slice and adhered in a fine sharp fluorescent line (under UV light) to the side of the dialysis tubing.
7 Reverse the polarity and electrophorese for a further 20 sec to detach the DNA from the wall of the tubing. Open one end of the dialysis bag and remove the DNA solution to a sterile corex tube or Eppendorf tube. There should be no fluorescence (DNA) left in the gel slice.
8 Add an equal volume of phenol/chloroform (50:50), equilibrated with TE buffer, vortex and then centrifuge for 1 min to separate the phases. Remove the upper aqueous phase having DNA. Back extraction of the phenol phase can be performed if isolating small amounts of DNA. Pool the aqueous phases.
9. Extract (deproteinize) the aqueous phases tube with an equal volume of a mixture of chl oroform-isoamyl alcohol (24:1). Remove the upper aqueous layer and depropeinize again until protin is seen at the interface.
10. **Important**: If the aqueous phase volume is more than 0.2 ml, extract with several volumes of n-butanol until the volume of the aqueous layer has been adequately reduced (i.e. until the desired DNA concentration has been achieved).

11. Precipitate the DNA by first, add a tenth of volume 2 to 3 M sodium or potassium acetate pH 5.5, mix thoroughly and then adding 2 to 3 ml of chilled (20^{0}C) absolute ethanol at, Centrifuge, wash the precipitate, dry and redissolve in sterile distilled water or TE buffer.

Note: This method is very quick and produces DNA of a quality comparable with that obtained using low melting temperature agarose. No further purification of the DNA by passage through an ion-exchange column etc. is necessary. In addition to excellent restriction and ligation, high specific activities by nick translation have been achieved.

Suggested readings

Birnboim H.C. and Doly J (1979). A rapid alkaline extraction procedure for screening recombinant plasmid DNA. Nucleic Acids Res. 7: 1513-1523.

Sambrook J., Fritsch F.F. and Maniatis T. (1989). Molecular cloning-A laboratory Manual, Cold Spring Harbor Laboratory Press, New York.

3

Transformation of Bacterial Cells

This is a procedure for transforming bacterial cells with circular DNA plasmids containing an origin of autonomous DNA replication and a selective marker (antibiotic resistance). Logarithmically growing cells are made permeable to DNA by incubation in $CaCl_2$ and the DNA is taken up from the surrounding buffer. This uptake is facilitated by a brief heat shock. Cells that have successfully taken up the DNA plasmid are then selected for growth on an agar plate containing the antibiotic. Cells that have not taken up the plasmid will not grow on these plates while transformants will form discrete bacterial colonies carrying plasmid. These can be further screened using minipreps and restricting the plasmid DNA using appropriate restriction enzymes if the transfer DNA plasmid is a chimeric plasmid DNA. The transformants are used to prepare plasmid DNA for sequencing etc.

Reagents

1. 2 x 10 ml aliquots of L-broth.
2. 1 x 100 ml aliquot of L-broth in one litre flask or 60 ml aliquot of broth in each of two 250 ml flasks.
3. 0.1M Calcium chloride (60 ml). Keep cool prior to use.
4. QB-agar-antibiotic plates.

i) Competent cells preparation

1. Set up 2 x 10 ml of the appropriate strain (JR 107, JR 109 or HB 101) in L-broth. Commonly the cells are stored in glycerol. Use a wire loop to inoculate. Grow at 37^{0}C overnight on the rotary incubator.
2. Innoculate 2 x 60 ml aliquots of L-broth with 1 ml of overnight grown culture. Grow at 37^{0}C on a fast shaker until the O.D. reaches a value of 0.2 at 650 nm on Spectrophotometer (This is equivalent to an O.D. of 0.1 at 600 nm for side arm flasks or an O.D. of 0.14 at 600 nm for cuvettes).
3. Place the cultures on ice and cool quickly. Transfer 100 ml of culture to a sterile 100 ml centrifuge tube. Spin down the cells at 5000 rpm at 5-10^{0}C for 15 min.

4. Decant off the supernatant and drain remaining drops by Pastuer pipette. Resuspend in 10 ml of cold 0.1M $CaCl_2$ soln. Resuspend gently by pipetting up and down. Keep as cool as possible. Add a further 40 ml of 0.1 M $CaCl_2$ soln. Leave in ice for 20 min.
5. Spin down the cells as in step 3. Decant off the supernatant. Resuspend the pellet in 5 ml of 0.1N $CaCl_2$ soln. STORE IN ICE for 30 min. cells are now called competent and can be used for transformation or stored in small aliquots (200 µl) quick freeze in liquid N_2 and store at -70°C for use later. Cells at this stage are fragile, therefore, handle cells gently.

ii) Transformation of competent cells and plating of cells

1. For frozen cells, first thaw on ice for 1-2 min. In a sterile Eppendorf tube, mix about 0.1 µg of plasmid DNA with 200 µl of competent cells.
2. Insubate on ice for 30 min.
3. Heat shocks the cells by transferring the tubes to 42°C for 2 min in a water bath.
4. Add 1 ml of LB medium and grow for 45 min at 37°C with shaking, to allow expression of the antibiotic resistance gene.
5. Use bent-end glass rod (sterilized with 70% ethanol and flaming) to spread cells over surface of LB-agar-antibiotic plates. Spread several different amounts (e.g. 50, 100, 150 & 200 µl) on separate plates to generate adequate numbers of well separated colonies. Store remaining cells at 4°C, if necessary.
6. Add 100 µl of untransformed bacterial cells to another plate with antibiotic as a negative control.
7. Incubate plastes at 37°C agar side down for 20 min to allow medium to dry.
8. Invert plates (agar on top) and grow overnight at 37°C. Colonies will show up next morning and can be processed further as desired.

Note:

1. To monitor the efficiency of the ligation and transformation steps, host cells should also be transformed with supercoiled plasmid as well as linear vector DNA (both phosphorylated and dephosphorylated) which has been subjected to ligation.
2. Bacterial strains DH5α, JM 109, JM 107 etc. are used as host strains for transforming pUC series of plasmids, pBluescript etc. HB 101 is normally used for pBR 322. Final concentrations of antibiotics in the plate should be 100 µg/ml of tetracycline or 30 µg/ml for kanamycin or 20 µg/ml for choramphenicol.

iii) Cloning of DNA fragment to plasmid DNA

Reagents

- 50 x TAE

Tris	242 g
Glacial acetic acid	57.1 ml
0.5M EDTA (pH 8.0)	100 ml
Make up the volume to	1000 ml.

Autoclave and use

- 5 M NaCl

NaCl	29.25 g
Water to	100 ml

Autoclave and use

3M Sodium acetate, pH 4.8

Protocol

a) Linearization of vector DNA

1. Take about 2 μg of plasmid vector DNA in about 16 μl of sterile water.
2. Add 2 μl of restriction buffer and 2 μl of restriction enzyme.
3. Keep at 37^{0}C for about 90 min.
4. Prepare 0.7% agarose gel in 1 x TAE.
5. After the restriction is over, load 1-2 μl of digested sample alongwith undigested sample to check whether the digestion is complete.
6. Run on agarose gel for about 30 min at 50V in 1xTAE buffer and check under UV light.
7. If restriction is complete, add an equal volume of saturated phenol: chloroform: isoamyl alcohol.
8. Mix the contents gently.
9. Spin for about 2 min.
10. Take the upper aqueous phase.
11. Add 1/10th volume of 3 M NaOAc, pH4.8.
12. Add 2 volumes of chilled absolute alcohol (EtOH).
13. Keep for about 1 hr at -70^{0}C.
14. After 1 hr centrifuge the DNA at high speed (12,000 rpm) for 15 minutes.

15. Wash the pellet with chilled 70% ethanol.
16. Dry the pellet and dissolve in 10 μl of sterile distilled H_2O.
17. To check the concentration and quality of linearized vector DNA, load 1 μl of vector DNA and fragment DNA on 0.7% agarose gel and run at 40V for 1 hr. View under UV and assess the quality of DNA.

b) Ligation of DNA fragment

Set up the ligation reaction on ice as follow:

1. Take 1 μl (0.1 μg) of the vector DNA (linearized), in a sterile Eppendorf tube.
2. Add 1 μl of equimolar amount of insert DNA (eluted fragment) and 5 μl of sterile distilled H_2O. To this add 1 μl of 10 X ligation buffer and 1 μl of 10mM ATP.
3. Briefly spin down the samples and add 1 μl of T4 DNA ligase (3-4 U/ μl).
4. Incubate at 12-15^{0}C overnight and use the mixture to transform competent cells of E.Coli (DH5α or XL-1 Blue).

Note : Set up two control reactions each containing the vector aone and fragment DNA alone, separately.

iv) Transformation of *E.Coli*

Protocol

1. To 100 μl of competent cells prepared earlier and add 5 μl of ligation mix and mix gently.
2. Keep the tubes in ice for 30 min.
3. Transfer the tubes to 42^{0}C in a water bath for 90-120 sec. (To give heat shock to the cells).
4. Quickly transfer tubes to an ice bath and chill for 2-5 min.
5. Add 1 ml of LB medium and incubate at 37^{0}C for 1 hr with gentle shaking.
6. Plate he cells on appropriate selective medium using different volumes (200 μl, 400 μl, 500 μl).
7. Incubate the plates invertedly overnight at 37^{0}C and check for the transformats next morning.

v) Blue/White colour screening

For blue/white colour screening with vectors capable of a complementation (pUC series, pBluescript, pGem series etc.), host strains carrying F like JM series, NM 522, DH5a are used. They carry lac Z DM15 and *lac* Iq on an F' episomes. The strains carrying F episome should always be maintained on minimal medium plates supplemented with thiamine-HCl. This selects for F episome which carries a nutritional requirement for growth (proline biosynthesis) and decreases the number of false positives. To use blue/white screening for recombinants and to maintain antibiotic selection, plate transformed cells (JM 109 or DH5a) on LB plates containing appropriate antibiotics, 0.1 M IPTG and 40 μg/ml X-gal. Incubate overnight at 37^{0}C.

Alternatively X-gal and IPTG can be spread (50 μl of 50 mg/ml X-gal and allow these components to absorb atleast 30 min at 37^{0}C prior to plating cells.

Suggested readings

Ausubel F.M., Brent R., Kingston R.E., Moore D.D., Seidman J.G., Smith J.A. and Struhi K. (1987). Current Protocols in Molecular Biology, John Wiley and Sons, Publications.

Davis L.G., Dibner M.D. and Batty J.F. (1986). Basic methods in Molecular Biology, Elsevier, New York.

Maniatis T., Fritsch F.F. and Sambrook J. (1988). Molecular cloning -A laboratory Manual, Cold Spring Harbor Laboratory Press, New York.

4

Preparation of Bacterophage λ DNA

Enterobacteria phage λ is a bacterial virus, or *bacteriophage*, that infects the bacterial species *Escherichia coli* (*E.coli*). It was discovered by Esther Lederberg in 1950. The wild type of this virus has a temperate life cycle that allows it to either reside within the *genome* of its host through *lysogeny* or enter into a *lytic* phase, during which it kills and lyses the cell to produce offspring. Lambda strains, mutated at specific sites, are unable to lysogenize cells; instead, they grow and enter the lytic cycle after superinfecting an already lysogenized cell.

The phage particle consists of a head (also known as a *capsid*), a tail, and tail fibers. The head contains the phage's double-strand linear *DNA* genome. During infection, the phage particle recognizes and binds to its host, *E. coli*, causing DNA in the head of the phage to be ejected through the tail into the cytoplasm of the bacterial cell. Usually, a "*lytic cycle*" ensues, where the lambda DNA is replicated and new phage particles are produced within the cell. This is followed by cell *lysis*, releasing the cell contents, including virions that have been assembled, into the environment. However, under certain conditions, the phage DNA may integrate itself into the host cell chromosome in the *lysogenic* pathway. In this state, the λ DNA is called a *prophage* and stays resident within the host's *genome* without apparent harm to the host. The host is termed a *lysogen* when a prophage is present. This prophage may enter the lytic cycle when the lysogen enters a stressed condition.

Protocol

i) Infection

1. Inoculate in duplicate 50 ml LB+10 mM $Mgcl_2$ + 0.2% maltose in 500 ml flast by a overnight grown culture of LE 392. *E.coli* strain. Grow for 2-3 hrs till the OD_{600} reaches ~0.4.
2. Calculate the cell concentration assuming that 1 $OD_{600} = 8x10^8$ cells/ml.
3. Pour into two autoclaved SS-34 rotor tubes and centrifuge at 4000 x g for 10 min at room temperature. Discard the supernatant.

4. Resuspend the bacterial cell pellet in 4 ml of SM buffer.
5. Add bacteriophage from the stock to a concentration of 5 x 10^8 phage and add 10 µl EMBL-3 phase stock given (4.6x10^7 phages/ ml) and mix rapidly.
6. Incubate at 37°C for 20 min with intermittent shaking.
7. Add infected aliquot 1 ml each to 125 ml of LB) + 10mM $MgSO_4$ + 0.2/mm maltose media in a 500 ml flask. Incubate at 37°C with very vigorous shaking till it is lysed. Complete lysis is judged by disappearance of silkiness.
8. Generally it takes 7-9 hrs. at least for complete lysis. For checking completeness of lysis, take 1 ml of the infected culture from flask. Add few drops of chloroform if it becomes clear proceed ahead if not continue incubation.
9. Add 2.5 ml chloroform to each flask and continue vigorous shaking for another 30 min. (chloroform causes the cells to burst if the infection is near completion.

ii) Purification of bacteriophage

1. Chill the culture to room temperature and add pancreatic DNase and RNase, both to a final concentration of 1µg/ml. Incubate for 30 min at room temperature.
2. Add solid sodium chloride to a final concentration of 1M (29.2g/500 ml of culture). Dissolve by swirling. Let it stand for 1 hr on ice.
3. Remove debris by centrifugation at 11,000 x g for 10 min at 4°C. Pool the supernatant in a clean flask. Repeat this step.
4. Add solid polyethylene glycol (PEG 6000) in small quantities with slow shaking to a final concentration of 10% w/v (i.e. 50 g/500 ml of supernatant.
5. Cool in ice water and keep it at 4°C preferably overnight.
6. Recover the precipitated bacteriophage particles by centrifugation at 11,000 x g for 10 min at 4°C. Discard the supernatant.
7. Resuspend the bacteriophage pellet in 7 ml SM buffer (l.e. 7 ml/500 ml supernatant).
8. Add equal volume (7 ml) of chloroform to the bacterlophage suspension and mix thoroughly. Centrifuge at 1600 x g for 15 min at 4°C. Recover the aqueous phase containing the bacteriophage.
9. Measure the volume of the supernatant and add 0.5 g/ml of solid cesium chloride. Mix gently. When the cesium chloride has dissolved, carefully layer the bacteriophage suspension onto cesium chloride step gradients that are prepared in Beckman SW 27 cellulose nitrate centrifuge tubes.

(Step gradients are made by layering carefully and sequentially solutions of decreasing density on top of one another. Make a mark on the outside of the tube opposite the position of the interface of two solutions and make step gradients using 1.45 g/ml and 1.50 g/ml cesium chloride solution).

10. Centrifuge using SW 27 rotor at 22,000 rpm for 2 hr at 4°C. A bluish band of bacteriophage particles should be visible at the Interface between the 1.45 and 1.50 g/ml layers.
11. Collect the band of bacteriophage particles using Siliconised Pasteur pipette. Store at 4°C in tightly capped tube.

iii) Extraction of bacteriophage DNA

1. Remove cesium chloride from the purified bacteriophage preparation by dialysis at room temperature of 1 hr against a 1000 fold volume of dialysis buffer.
2. Change the dialysis buffer and dialyse for an additional hour.
3. Transfer the bacteriophage suspension into a centrifuge tube.
4. Add EDTA to give a final concentration of 20 mM.
5. Add pronase to a final concentration of 0.5 mg/ml, or add proteinase K to final concentration of 50 mg/ml.
6. Add SDS (20%) to a final concentration of 0.5%. Mix by inverting the tube several times.
7. Add an equal volume of equilibrated phenol. Mtx by inverting the tube several times. Centrifuge at 1600 x g for 5 min at room temperature. Transfer the aqueous phase to a clean tube.
8. Extract the aquous phase once with a 50:50 mixture of equilibrated phenol and chloroform. Extract once again with an equal volume of chloroform.
9. Precipitate the DNA by adding a tenth volume of 3M sodium or potassium acetate pH 5.5, mixing thoroughly and then adding 3 ml of absolute alcohol at -20°C Centrifuge, wash and pellet with 70% ethanol and then dissolve DNA pellet in 50 μl of TE buffer. Store at -4°C for further use.

Suggested readings

Hohn B. (1979). *In vitro* packaging of lambda phage and cosmid DNA. In methods of Recombinant DNA, Vol. 68, Wu ed., Academic Press.

Mandal, M. and Higa, A. (1970). Calcium-dependent bacteriophage DNA infection. J. Mol. Biol., 53: 159-162.

5

Isolaton and Purification of Genomic DNA from Plants

There are various methods available for isolation of high molecular weight genomic DNA from plants. The basic steps involved in all these methods are removal of cell wall, nuclear membrane, proteins, lipids, carbohydrates, cell debris, RNA without affecting the integrity of DNA. The two major factors that affect DNA integrity are action of endogenous nucleases and impact of mechanical shearing.To reduce endogenous nuclease activity the plant tissue is frozen in liquid nitrogen and ground to fine powder. The ground tissue is thawed in presence of extraction buffer containing detergent and high concentration of EDTA which helps to chelate the divalent cations. Mechanical shearing problem can be taken care of by avoiding harsh treatment of lysate. Here a simple method based on Dellaporta *et al* (1983) for isolation of intact DNA from plant material is described.

Reagents

- DNA Extraction Buffer
- 100 mM Tris-Cl (pH 8.0); 20 mM EDTA (pH 8.0); 1.4 M NaCl;
- 2.0% (W/V) CTAB; 0.2% 2-Mercaptoethanol
- Chloroform-Isoamylalcohol (24 : 1)
- TE Buffer: 10 mM Tris (pH 8.0); 10 mM EDTA

i) Isolation of genomic DNA from plants

Protocol

1. Weigh 15 g of leaf tissue (etiolated), freeze quickly in liquid nitrogen and grind to a fine powder in a 3" mortar and pestle. It is important not to let the tissue thaw once frozen until buffer is added.
2. Add 75 ml of DNA extraction buffer. For maximum yield, the cells are further broken by grinding the mixture at a low setting (about 3) with a polytron; however, this step is optional.

3. Transfer to a 250 ml flask, add 5 ml of 20% SDS, mix thoroughly by shaking vigorously for 20 min, then incubate tubes at 65^{0}C for 10 min.
4. Add 50 ml of 5 M potassium acetate. Shake flask vigorously and incubate at 0^{0}C for 20 min (keep on ice). Most proteins and polyseccharides are removed as a complex with the insoluble potassium dodecyl sulfate precipitate.
5. Spin tubes at 25,000 g for 20 min. Pour supernatant through a Mira cloth filter plantsmin.
6. Pellet DNA at 20,000 g for 15 min. gently pour off supernatant and dry pellets by inverting the tubes on paper towels for 10 min.
7. Redissolve DNA pellets in 3.5 ml buffer of 50 mM Tris, 10 mM EDTA (pH 8.0). Transfer the solution to an Eppendorf tube. Spin the tube in Eppendorf centrifuge for 10 min to remove insoluble debris.
8. Transfer the supernatant to a new Eppendorf tube and add 375 µl of 3 M sodium acetate and 2.5 ml isopropanol. Mix well and centrifuge for 1 min in a Eppendorf centrifuge to pellet DNA. Wash pellet with 80% ethanol, dry using nitrogen gas stream or under vacuum and redissolve in 500 µl of TE buffer.

Precipitation with 3M sodium acetate using relatively small amounts of isopropanol (about 0.6 vol.) has been reported to separate high molecular weight DNA from polysaccharides. Sodium acetate also yields a tight fibrous precipitate that is easily washed and dried.

ii) Purification of DNA by ultracentrifugation

1. Add 4.5 g of CsCI and 200 µl Ethidium bromide solution.
2. Pellet the precipitate, if any, and transfer the supernatant using syringe (5 ml) and needle (18 gauges) to polyallomer quick seal centrifuge tubes.
3. Heat-seal the tube after proper balancing.
4. Spin the tube at 60,000 rpm in Ti 65 rotor for 18 hrs at 20ºC.
5. Visualize DNA band using UV torch and take out the DNA band with the help of syringe (2 ml) and needle (18 gauges) and transfer to Eppendorf tube.
6. Add equal volume of butanol saturated with TE to remove Ethidium bromide.
7. Mix by gentle inversion and spin for 2 min.
8. Carefully remove the upper phase and discard & repeat till the lower phase becomes colourless.

9. Add an equal volume of isoproponol, mix by inverting the tubes several times, keep at -20^{0}C for 1 hr and Pellet, DNA by centrifugation.
10. Dissolve DNA Pellet in 500 µl of TE buffer and the estimate concentration by spectrometer.
11. Quality and quantity of DNA

The procedure for isolation of DNA described above generally gives excellent DNA with no contaminating polysaccharides, RNA or protein. Nevertheless, it is essential to check the purity of each preparation before further studies. A good DNA preparation generally exhibits the following spectral properties z:

$A_{230} < 0.10$; $A_{230/260} \sim 0.55$; $_{A280/260} \sim 1.8$

A conversion factor 50 is used to convert OD to concentration in µl/ml.

DNA Concentration (µg/µl) = OD260 x 50 dilution factor/1000. If ratio OD260/OD280 is 1.8~2.0 then DNA is pure. If ratio is less than 1.8, there may be protein (reprecipitate the DNA). A ratio higher than 2.0 indicates the sample may be contaminated with chloroform or phenol and reprecipitate with ethanol.

iii) Alternative method of isolation and purification of genomic DNA

Isolation of high molecular weight DNA that is suitable for digestion with restriction endonuclease is an essential requirement for confirming transgene integration through polymerase chain reaction or by southern blot hybridization. The plant tissues are notoriously difficult material for DNA isolation due to the presence of various secondary plant products. A protocol that works with one plant group may fail with others. Consequently, a number of DNA isolation methods have been developed for different target plant groups. Two methods developed originally by Dellaporta *et al.* (1983) and Murray and Thompson (1980) respectively, are being commonly used with certain modifications for isolating high molecular weight DNA from small amounts of tissue. In our hands a modified Murray and Thompson protocol, as described below has worked equally well in a number of plant genera including rice, wheat, chickpea, *Nicotiana* and *Brassica.*

Reagents

- Extraction buffer: 2% (w/v) CTAB, 100 mM Tris-Cl, pH 8.0, 1.4 M NaCl, 20mM EDTA, 0.2% mercaptoehtanol.
- TE buffer: 10 mM Tris, 1mM EDTA, pH 8.0
- Isopropanol
- Ethanol
- Chloroform

- Isoamyl alcohol
- RNase
- Agarose
- Ethidium bromide

Protocol

a) DNA isolation

1. Grind 0.5 g of leaf material in liquid nitrogen to fine powder using pre-chilled mortar and pestle.
2. Transfer the powder to a 15 ml polypropylene centrifuge tube containing 5 ml of pre-warmed extraction buffer. Use spatula to disperse the material completely.
3. Incubate samples at 60°C for 30 min with occasional mixing by gentle swirling.
4. Add 3 ml chloroform isoamyl alcohol and mix by inversion to emulsify.
5. Spin at 15,000 rpm for 10 min at room temperature.
6. Remove the aqueous phase with a wide-bore pipette, transfer to a clean tube, ass 2/3 volume of isopropanol and mix by quick gentle inversion.
7. Spool DNA using a bent Pasteure pipette and transfer to another tube. Alternatively, if the DNA appears flocculent, centrifuge at 5000 rpm for 2 min and gently pour off the supernatant.
8. Wash the DNA pellet in 70% ethanol (5-10 ml) for 20 min.
9. Dry the pellet and dissolve in 500 µl TE buffer.

b) Purification of DNA

Major contaminants in crude DNA preparation are RNA, protein and polysaccharides. Inclusion of CTAB in DNA extraction buffer helps in elimination of polysaccharides from DNA preparations to a large extent. The RNA is removed by treating the sample with RNase. Proteins including RNase can be removed by treatment with proteinase K. Extraction with phenol: chloroform following RNase treatment is also employed for eliminating RNA and most of proteins. Given below is the protocol of DNA purification routinely were used in our laboratory.

1. Dissolve 10 mg RNase in 1 ml of 10mM Tris-HCl (pH 8.0) and 15 mM NaCl in an 1.5 ml microfuge tube.
2. Heat the tube in boiling water bath at 100°C for 15 min to denature contaminating DNase.

3. Cool slowly at room temperature.
4. Add RNase @ 50 pg/ml to DNA sample and incubate at 37^{0}C for one hour.
5. Add equal volume of phenol: chloroform (1:1) and mix.
6. Spin at 11000 rpm for 2 min at room temperature, take out the aqueous phase and transfer to a fresh microfuge tube.
7. Extract twice with equal volume of chloroform: isoamylalchohol (24:1), centrifuge and take out the aqueous phase.
8. Add 0.1 volume of 3M sodium acetate (pH 4.8) and mix properly.
9. Add 2.5 times absolute alcohol, mix by quick gentle.
10. Pellet the DNA by centrifugation at 10,000 rpm for 5 min in a microfuge.
11. Decant supernatant carefully, wash the pellet with 70% cold ethanol, air dry and dissolve DNA in 50-100 µl TE buffer and store at 4^{0}C.

iv) Restriction of genomic DNA

A. Reagents

- 10 x TBE buffer
- 0.89 M Tris; 0.89 M Boric acid; 0.02 M EDTA, Ethidium bromide (10 mg/ml)
- 10 x Gel loading buffer
- 0.25% Bromophenol blue; 0.25% Xylene cyanol; 25% Ficoll (type 400) in water;
- 5M Ammonium acetate
- 0.5 M EDTA, pH 8.0
- BSA (1 mg/ml)

B. Protocol

Restriction

1. Add the components of the reaction mixture to an Eppendorf tube in the following order.

Water	130 µl
10 x TBE buffer	20 µl
BSA	20 µl
DNA	20 µl (10 µg DNA)
Restriction enzyme(5 units/µlDNA)	10 µl (50 units)

2. Mix by gentle tapping and incubate at 37°C overnight.
3. Add 8 μl of 0.5 m EDTA to stop the reaction.
4. Add 2/3 volume of 5M ammonium acetate and mix.
5. Precipitate DNA by adding two volumes of ethanol.
6. Pellet DNA and wash in 70% ethanol.
7. Dry and dissolve in 30 – 40 μl of TE.
8. Store at 4°C

v) Agarose gel electrophoresis of DNA

Solutions

1. 10 x TBE

Tris base	= 108 g
Boric acid	= 55 g
0.5 M EDTA pH 8.0	= 40 ml
Water to 1 litre	

2. 10 x Gel loading dye

Glycerol	= 5 ml
10 x TBE	= 1 ml
Bromophenol blue (Saturated)	= 1 ml
Xylene cyanol 10%	= 1 ml
Mix well autoclave and store in aliquots	

3. Ethidium bromide in sterile water = 10 mg/ml

Protocol

1. Weigh appropriate amount of agarose for making 0.7% gel.
2. Add appropriate amount of 1 x TBE and boil to melt the agarose.
3. Cool the melted agarose solution to -50°C, add ethidium bromide (μl / ml final concentration) and pour to the gel casting tray with comb in position.
4. Allow the gel to set and remove the comb and keep the gel tray in the electrophoretic tank. Pour 1 x TBE just enough to cover the gel.
5. Prepare the DNA sample by adding 1/10 volume of 10 x loading dye and load onto the wells.
6. Load a known amount of uncut phage DNA as control to an adjacent well.

7. Run the gel at 50 V for 1 hr.
8. Visualise the DNA band under UV light in a transilluminatior. Presence of single intact band at the corresponding position to phage DNA indicates high molecular weight of the isolated DNA.

Suggested readings

Dellaporta S., Wood J and Hicks J.B. (1983). A plant DNA mini preparation Version II. Plant Mol. Bio. Rep 1: 19-21.

Doyle, J.J. and J.L. Doyle (1990). Isolation of plant DNA from fresh tissue. Focus 12: 13-15.

Murray M.G. and Thompson W.F. (1980). Rapid isolation of high molecular weight plant DNA. Nucl Acids Res. 8: 4321.

Seema Khitha, Rekha Kansal and Koundal K.R. (1995). Isolation and characterization of genomic DNA of Chickpea (*Cicer arietinum* L.). Plant Physiology and Biochemistry 22 (1): 1–5.

6

Isolation of Total RNA and *in vitro* Translation of Poly (A)$^+$ RNA

RNA is transcriptional product of DNA. The total RNA includes three classes, rRNA, tRNA and mRNA. The ribonucleic acids exists as ribonucleo protein particles in intact cells. The ribonucleo protein complex is dissociated by SDS in to RNA and protein deproutinised by phenol and the free RNA left in the aqueous solution is precipitated in cold by adding alcohol. The RNA is more susceptible to degradation than DNA, due to the ability of the 2' hydroxyl groups adjacent to the phosphodiester linkages in RNA to act intermolecular nucleophiles in both base and enzyme catalysed hydrolysis. Whereas deoxyribonucleases (Dnases) reuire metal ions for activity and therefore can be inactivated with chelating agents e.g. EDTA, many ribonucleases (Rnases), bypass the need for metal ions by taking advantages of the 2' hydroxyl group as a reactive species. The RNase single srand specific endonuclease that is resistant to metal chelating agents can survive prolonged boiling or autoclaving. However, RNase A type enzymes rely on active site histidine residues for catalyzing activity and consequently, these enzymes can be inactivated by the histidine specific alkalizing agent diethyl pyrocarbonate (DEPC). The Contaminating Rnases can inactivated by baking glassware at 180^{0}C or higher for several hours. The DEPC will destroy polycarbonate or polystyrene materials (e.g. electrophoresis tank) which should instead be decontaminated by soaking in 3% hydrogen peroxide for 10 minutes. Remove peroxide by extensively rinsing with DEPC treated & autoclaved water.

Various procedures are used for extraction and purification of total RNA and mRNA from different sources. The method selected largely depends on the source material and the experimental use of RNA extracted. The most commonly method (Chirgwin *et al.*, 1979) is described below for isolation of total RNA from plant material.

i) Isolation of total RNA from plants

Reagents

Solution A: 4M Guanidine isothiocyanate solution (GTC)

50 g GTC; 2.5 ml 1 M Sod Citrate pH 7.1; 0.7 ml 2-mercaptoethanol. Adjust pH to 7.0 with citric acid; make up the volume to 100 ml. Do not autoclave or filter. Use autoclaved Mill-Q water for every solution.

Solution B: CsCl cushion solution

48g CsCl, 1 ml of 0.5M EDTA (pH 7.0); volume to 50 ml (autoclave)

Solution C: Resuspension buffer (50 ml)

IM Tris-CI (pH 7.5) 0.5 ml; 0.5M EDTA (pH 8.0) 1.0 ml; 20% SDS 0.125 ml; 5M NaCI, 3.5 ml; Sterile H_2O, 44.975 ml. Autoclave, dispense in Aliquots and store at -20°C.

Protocol

1. Homogenize 1 g material (Seed or embryos) in 10 ml GTC solution in cold for 20 sec each x 4 times using polytron tissue homogenizer and pass through 4-layers of cheese cloth.
2. Centrifuge for 15 min at 10,000 rpm at room temperature. Take out the supernatant carefully. Spin at 27,000 rpm for 23 hrs at 25°C. Remove GTC layer carefully without disturbing pellet. Invert the tube on tissue paper and remove any traces of GTC.
3. Cut the tube 1 cm from bottom and slowly add RNA suspension buffer (100 µl - 150 µl each time). Dissolve the pellet and transfer the material to the Eppoendorf tube. Resuspend RNA in 600-700 µl of Solution C and heat at 60°C in a water bath for 60 min vortexing in between.
4. Keep the tubes now in ice for sometime and centrifuge at 4°C for 10-15 min in Eppendorf centrifuge.
5. Extract the supernatants with equal volume of phenol: chloroform (1:1). in 1 ml volume with 1 ml of phenol:chloroform (1:1) mixture.
6. Take the aqueous phase and add 2 vol of 100% cold ethanol and keep at -20°C overnight (Do not use -70°C for precipitation).
7. Pellet RNA and wash the pellets with 70% ethanol dry down the pellet and dissolve in sterile water. Estimate RNA yield at A_{260} by using a relationship 1OD = 40 µg RNA.

ii) Aquarose gel lectrophoresis of total RNA

Reagents

1. 10 x MOPS solution (pH 7.0), 41.86 g MOPS; 6.8g NaOAC (Sodium acetate); 3.362g EDTA. Adjust pH 5.5 to 7.0 with sulphuric acid make up to 1000 ml and autoclave. Solution turns to yellow after autoclaving.

2. Loading buffer (5X): Glycerol 10 ml; 0.5M EDTA; 40 µl Bromophenol blue 40 mg; Xylene cresol 40 mg; H_2O to 20 ml.

Note: Soak the gel casting tray, comb, buffer containing cells in proteinase K (25 mg/1000 ml) and 0.1% SDS in sterile H_2O overnight. Rinse then properly and use for pouring gel.

Protocol

i) Preparation of agarose gel

To 1.2 g agarose add 86.6 ml sterile water and boil to dissolve. Allow it to cool to 55°C and then add 12 ml of 10 X MOPS and 6.5 ml formaldehyde (37%). Mix by swirling. Position sample well former (comb) and pour agarose solution on plate to form gel.

ii) Preparation of sample

Take 4 µl RNA (if concentrated use less and make up with water). 1.56 µl, 10 X MOPS; 3.12 µl Formaldehyde; 8.88 µl deionised formamide; Heat at 60°C for 7.5 min. Cool in ice for 5 min. Add 4 µl 5 X loading buffer.

iii) Running gel

Place slab gel into electrophoresis tank and use 1 X MOPS as running buffer.

a) Load the sample on gel and run at 10V for overnight (40V for big gel).
b) Stain the gel with ethidium bromide (1 µg/ml of sterile water). Shake for 10 min.
c) Destain with several changes in sterile water. Photograph gel on transilluminator. Total RNA contains abundant amounts of 28S RNA (5kb) and 18S RNA (~2 kb) that can be used as molecular weight standard.

iii) Isolation of Poly $(A)^+$ RNA from total RNA

Isolation of mRNA is often required to analyze gene expression. However, mRNA makes up only 1-5% of the total cellular RNA fraction. The vast majority of the RNA consists of rihbosomal (rRNA) and transfer RNA (tRNA). The actual amount of mRNA present depends on the cell type and the physiological state of the cells when they are lysed. Due to the low abundance of mRNAs in the total population of cellular RNA, removal of rRNA and tRNA from a total RNA preparation results in a dramatic increase in the relative amount of mRNA.

The isolation of mRNA capitalizes on the fact that most eukaryotic mRNAs end in a homopolymer of 20-300 adenosine residues, commonly called the 'poly (A^+) tail'. Poly (A^+) tails are added to the RNA transcript in the nucleus following transcription. In contrast, rRNA and tRNAs are not polyadenylated. The poly (A^+) tail thus provides a unique feature for the isolation of mRNA. Fractionation of mRNA away from rRNA and tRNA is achieved by hybridization in high salt of the polyadenylated mRNAs to oligo dT primers coupled to a solid support consisting of cellulose. The RNA species lacking poly (A^+) tails fail to bind, or are weakly bound to the oligo dT, & can be removed by washing the matrix with a low salt buffer. The mRNA is then eluted from the oligo-dT with a zero salt buffer.

Length of poly (A^+) tails varies approximately from 20 bases to 300 bases.Molecules with as few as eight adenylate residues bind to the oligo-dT assuring that mRNA species with short poly (A^+) tails would not be preferentially lost during purification. In general poly (A^+) RNA yield is 1% total RNA.

The quality of the isolated mRNA is defined by the presence of intact, full length mRNA as well as the degree to which contaiminating ribosomal RNA has been removed. These characteristics may be checked by native agarose gel electrophoresis of the purified mRNA. Usually one round of selection is sufficient for many applications (eg. Northern analysis and Ribonuclease protection assays), the presence of residual ribosomal RNA after a single round of purification is problematic for cDNA synthesis for library construction. The ribosomal RNA generates ribosomal cDNA which increases the effort required to isolate rare transcripts from cDNA libraries:

Reagents

1. Oligo (dT) cellulose matrix
2. 0.1M NaOH with 5mM EDTA, pH 8.0
3. Loading buffer (A)
   40 mM Tris-CI pH 7.4; 1M NaCI; 1mM EDTA; autoclave. Add SDS to 0.1%
4. Loading buffer (B)
   20 mM Tris-CI pH 7.4; 0.1M NaCI; 1mN EDTA; autoclave.
5. Elution buffer (C)
   10 mN Tris-CI pH 7.4, 1mM EDTA; autoclave.

Note: Autoclaved buffer, column and tubing are to be used. Add SDS to 0.1% (0.5ml of 20% SDS/100 ml) in loading buffer. SDS is not autoclavable.

6. 5M NaCl.

Protocol

1. Prepare column by pouring approximately 0.5 ml oligo (dT) cellulose slurry suspended in loading buffer B into a 1 ml disposable plastic syringe. Wash

with 3 ml of 0.1M NaOH with 5 mM EDTA. Wash column with H_2O until pH of effluent is less than 8. Wash column with 5 ml of loading buffer A to equilibrate the column.

2. Take about 5-10 mg of total RNA in 500 µl of H_2O. Heat to 65°C in water bath for 5 min. Add equal volume of loading buffer. A pre-warmed at 65°C. Mix and cool to 4°C for 2 min.
3. Apply this 1ml sample to column. Collect eluate. Heat eluate to 65°C for 5 min. Cool to 4°C for 2 min and reapply to column.
4. Wash with 5 ml of loading buffer B. This eluate is not poly (A^+) RNA and can be discarded.
5. Now elute with elution buffer C and collect 10 drop fractions.
6. Add 5M NaCl to make it 0.2M NaCl and then add 2.5 vol of absolute ethanol and keep at -20°C for overnight.
7. Centrifuge RNA at 25,000 x g for 2 hrs at 4°C and remove the supernatant. Rinse pellet in 500 µl of 80% ethanol, spin in microcentrifuge. Dry pellet under vacuum.
8. Dissolve the pellet in appropriate amount of sterile water. Check absorbance at 260 nm. App. 2-4 µg RNA can be run on 1% formaldehyde agarose gel to check the quality of poly (A^+) RNA isolated.

iv) *In vitro* Protein translation

Cell-free protein synthesizing systems have been standard tools for the *in vitro* expression of proteins from cloned genes. In vitro systems allow:

i) Use of a defined template to director proten systhesis,

ii) Specific labeling of gene products so that individual proteins can be monitored in complex reaction mixtures and

iii) Rapid means for protein analysis as template need not necessarily be cloned prior to expression.

Protein products of these systems can be monitored by two ways.

(i) The labeled gene product can be identified by autoradiography following SDS-PAGE and (ii) the non-radioactive labelling method exists in which if the antibodies to the protein are available the immunopreciitation or immunoblotting directly from *in vitro* translation reaction can be done.

Rabbit reticulocyte lysate reaction system for *in vitro* translation

Reagents

1. Appropriate Rabbit Reticulocyte Lysate system (Promega)

2. RNAasin-Ribonuclease Inhibitor
3. Radiolabelled amino acid
4. Nuclease free water

Final Concentration of RRL components in a 50 µl translation reaction

Creatine phosphate	=	10 mM
Creatinine phosphokinase	=	50 µl/ml
DTT	=	2 mM
Calf liver tRNA	=	50 µg/µl
Potassum acetate	=	79 mM
Magnesium	=	0.5 mM
Acetate Hemin	=	0.02 mM

Protocol

1. Heat the template mRNA at 67°C for 10 min and immediately cool on ice. (This increases the efficiency of translation especially of GC rich mRNA, destroying local regions of secondary structure).
2. Assemble the reaction as follows:

Rabbit Reticulocyte lysate	=	35 µl
Nuclease free water	=	7 µl
RNAasin Ribonuclease Inhibitor (at 40 U/µl)	=	1 µl
1mM Amino Acid Mixture (minus methionine)	=	1 µl
[^{35}S] methionine (1200 Ci/mmol at 10 mCi/ml)	=	4 µl
RNA substrate in water	=	2 µl
Final volume	=	50 µl
Final [^{35}S] methionine conc	=	0.8 mCi/ml

Note: For the positive control reaction use 2 µl of the control RNA supplied.

3. Incubate the reaction at 30°C for 60 min.
4. Analyse the result of translation by autoradiography following SDS - PAGE

v) Electrophoresis of translation products

The most versatile method used for analysis of *in vitro* translation products is separation on SDS-polyacrylamide gel electrophoresis followed by flurography to detect the synthesized products. The electrophoresis is performed in the presence of 0.1% SDS and a discontinuous buffer system. The separating gel at a concentration of 15% acrylamide gives good separation of polypeptide mixtures between 20,000 and 1,00,000 molecular weight.

Reagents

1. Separating gel 4 x Buffer (per 100 ml)

   Tris-base = 18.17 g

   10% SDS = 4 ml

   Adjust to pH 8.8 with HCl and add H_2O to a final volume of 100 ml. Store at room temperature.

2. 30% acrylamide solution (per 100 ml)

   Acrylamide = 30 g

   Bisacrylamide = 0.8 g

   Add H_2O to a final volume of 100 ml. Store at 4°C

3. Stacking gel 4 x buffer (per 100 ml)

   Tris-base = 6.06 g

   10% SDS = 4 ml

   Adjust to pH 6.8 with 1N HCl and add H_2O to a final volume of 100 ml. Store at room temperature.

4. Loading buffer 2 x (per 10ml)

   Glycerol = 2.0 ml

   10% SDS = 2.0 ml

   Bromophenol blue = 0.25 mg

   Stacking gel 4 x buffer = 2.5 ml

   β-mercaptoethanol = 50 μl

   Add H_2O to a final volume of 10 ml. Store at room temperature

5. Running 10 x Buffer (per litre)

   Tris-base = 30 g

   Glycine = 144 g

   10% SDS = 100 g

6. Fixer : 30% Methanol + 10% Acetic acid

7. Enhancer

Protocol

1. Preperation of 15% gel

   Mix the following from the stock to prepare separating gel.

   H_2O = 18 ml

   30% Acrylamide stock solution = 37 ml

   Separating gel 4 x buffer = 19 ml

   10% ammonium per sulphate solution = 1.12 ml

   TEMED = 50 µl

   (This will be sufficient for 2 gels)

   Pour the gel mix into the assembled glass plates leaving sufficient space at the top for the stacking gel to be added later. Gently over lay the gel mix with 0.1% SDS

2. After polymerization (15-30 min), remove the overlay and rinse the surface of the gel with water.

3. Prepare a stacking gel mix as follows:

   H_2O = 10 ml

   30% Acrylamide = 3 ml

   Stocking gel 4 x buffer = 4.44 ml

   10% ammonium per sulphate = 280 µl

   TEMED = 50 µl

   Fill the remaining space on the top of the separating gel with stacking gel mix and insert the comb immediately.

4. After stacking gel is polymerized (30-45 min.) remove the comb and rinse the wells to remove any umpolymerised acrylamide.

5. Attach the gel assembly to the electrophoresis tank.

6. To the *in vitro* translateod product (10 µl) add 20 µl of 2x loading buffer and heat the sample at 100°C for 2.5 min. and load the sample onto the gel.

7. Fill the upper and lower tank with 1 x running buffer and run the electrophoresis at a constant current of 15 mA in the separating gel and 30 mA in the separating gel. The electrophoresis is performed until the bromophenol blue dye reaches the bottom of the gel.

8. The gel after the end of the run is first fixed in 30% methanol and 10% acetic acid for 30 min.

9. The gel is then dipped in Enhancer solution for 30 min for 1 hr.

10. Place the gels on a sheet of Whatmann 2MM filter paper and cover it with Saran Wrap and dry under vacuum.
11. Gel is then exposed to X-ray film for appropriate time and develops. Analyse the protein bands.

vi) Autoradiography

Radioactive nucleic acids and proteins may be detected by autoradiography or by liquid scintillation spectroscopy. Autoradiographic methods are considered superior as they are more sensitive and give higher resolution. Readymade developer and fixer are available however, these can be prepared as follows:

A. Requirements

1. X-ray film developer

   Water at 52°C = 500 ml

   Metol = 1 g

   Sodium sulphite = 75 g

   Hydroquinone = 9 g

   Sodium carbonate monohydrate = 25 g

   Dissolve in order and make up to 1 litre with cold water.
2. Fixer

   Sodium thiosulphate = 80 g

   Potassium metabisulphite = 1 g

   Dissolve in order in 250ml and filter
3. Stop bath 2% acetic acid
4. X-ray film
5. Cassette with intensify screen

Protocol

1. Saran wrap the material to be autographed
2. Place pieces of tape marked with radioactive ink at several locations around the edge of sample.

   (Radioactive ink is made by mixing a small amount ^{32}P with a water proof black ink. It is convenient to make the ink in three grades:

   Very hot = 2000 cps on a minimonitor

   Hot = 500-2000 cps on a minimonitor

   Cool = 50-500 cps on a minimonitor

3. When the ink is dry, wrap the sample and backing sheet in Saran Wrap
4. In a dark room place the sample in a X-ray film holder and cover it with a X-ray film (use the screen if required. The sensitivity of the film may be increased if intensifying screen is used.
5. Close the cassette and wrap it in black bags
6. Keep at -70°C for several hours to several days (this depends on the radioactivity).
7. Remove the film after exposing and develop it is developer for 5 min. in dark.
8. Wash the film in stop bath for 1 min. and fix the film in fixer for 5 min. in dark.
9. Wash under running water for 30 min.
10. Dry the film at room temperature.

Suggested readings

Chirgwin J.M., Przybyla A.E., MacDonald R.J. and Ratter W.J. (1979). Isolation of biologically active ribonucleic acid from sources enriched in ribonucleases. Biochemistry 18: 5294-99.

Chomczynski P. and Saach N (1987). Single step method of RNA isolation by acid guanidinium thiosyanate phenol- choloroform extraction. Anal. Biochem. 162: 146-159.

Koundal K.R., Siddiq E.L. and Mehta S.L. (1977). Changes in soluble proteins and isoenzymes in rice (*Oryza sativa*) during germination. Indian Journal of Experimental Biology 15: 388–390.

Laemmli U.K. (1970). Cleavage of structural proteins during the assembly of the head of bacteriophage T4. Nature 227: 680.

7

Isolation and Analysis of Polyribosomes and Ribosomal Proteins

Polysomes consist of two or more ribosome traversing a strand of mRNA, translating the nucleotide sequence into the corresponding amino acid sequence. Usually at any time, many ribosomes are translating an mRNA simultaneously forming a structure called a polyribosome or polysomes. One reason for isolating polyribosomes is to resolve them on sucrose gradients to determine the extent of ribosome loading. The second main reason is to provide a source of mRNA for *in vitro* translation. The major factor influencing the number of ribosomes per mRNA molecule is RNase-an enzyme that is difficult to inhibit, and many polysome isolation protocols are designed solely to prevent its action. The RNase activity initially causes the conversion of large polysomes into small ones, and, only at later stages, the monosomes accumulate. Disaggregation can also result from ribosome run-off, which leads to the accumulation of monosomes and only a slight shift from large to smaller polysomes. There are many protocols available in literature for the isolation of polyribosomes by differential centrifugation. In general the method involves, layering of post mitochondrial supernatant of a cell lysate or homogenate on to a 1M molar sucrose cushion in a neutral or alkaline buffer (pH 7.6-8.5) and centrifugation at 300-400K g for 2-3 hours to pellet the polysomes. The polysome pellets are then resuspended in TE buffer for further analysis.

i) Isolation of polyribosomes

Reagents

1. Buffer A, pH 8.5 (1000 ml)

Trisbase	=	0.2 M (24.23g)
Sucrose	=	0.2 M (68.46g)
KCl	=	60 mM (4.47 g)
$MgCl_2$	=	30 mM (6.100g)
		Adjust pH 8.5 with HCl and autoclave
DTT	=	1mM (0.154g), (added after autoclaving)

2. Buffer B. pH 8.5, (1000 ml)

Trisbase	=	40 mM (4.844 g)
KCl	=	20 mM (1.49 g)
$MgCl_2$	=	10 mM (2.033 g)
	=	Adjust pH 8.5 with HCl and autoclave

3. 70% sucrose in Buffer B

   Sterilize sucrose (Rnase free) after weighing separately and then add sterilized buffer B to it.

4. Triton X – 100 – 20% in buffer A

   Sterilize sucrose (Rnase free) after weighing separately and then add sterilized buffer B to it.

Protocol

1. Weigh 20g tissue. Add 60 ml homogenisation buffer A. Homogenise with Polytron with speed setting at 8 for 30 sec.
2. Add 4% Triton (final concentration) using 20% triton x-100 and stir for 15 min. Filter through mira cloth.
3. Centrifuge at 2,000 rpm in SS - 34 for 10 min at 4°C in RC5B centrifuge.
4. Transfer supernatant to another set of 50ml centrifuge tubes and centrifuge at 16,000 rpm for 10 min at 4°C.
5. Remove the supernatant and layer it over 10 ml 70% sucrose pad in buffer B in ultracentrifuge tubes. Use a syringe and needle to ensure the removal of all air bubbles from the tubes. Cap the tubes properly, balance them (the tubes must be completely filled, use buffer A if necessary).
6. Centrifuge at 55,000 rpm for 2 hrs at 4°C using 70Ti rotor.
7. Allow the centrifuge to stop without brake.
8. Discard supernatant, rinse quickly and gently with sterile water and wipe tubes with tissue, taking care not to touch the pellet.
9. The tubes with pellet can be frozen in liquid nitrogen and stored at -70°C if it is to be used later.
10. Quantitative estimation of polyribosomes

    Remove polysomal pellet from -70°C cabinet melt quickly in hands and place in ice bucket. Dispense each pellet in 500 μl of buffer (50 mM Tris, 30 mM KCl, pH 7.4) using a glass rod. Mix 5 μl polysome + 995 μl H_2O and take OD260 mm. Use the conversion factor 1 mg polyribosomes = 10 OD_{260}

e.g.: If OD at 260 = 0.48

Polysome concentration = 48 μg / 5μl = 9.6 μg / μl = 9.6 mg/ml.

ii) Isolation of Poly $(A)^+$ rich RNA from polysomes.

The messenger RNA (mRNA) accounts only about (2-5% of the total RNA and the rest are tRNA and rRNA. The messenger RNA carries a stretch of polyadenylatetail at its 3' end. It may vary in length ranging 3-200 adenyl residues. Under high ionic conditions poly (A) tail of mRNA forms a duplex with the complement sequences such as poly (U) or oligo (dT) while tRNA and rRNA do not form any duplex since they lack poly (A) tail. When the ionic strength is lowered the hydrogen binding between poly (A) and poly (U) or oligo (dT) is broken thus releasing the poly (A) containing molecules. This is the basis for the development of affinity chromatography to isolate the mRNA from the total RNA.

Reagents

Buffer-I

1. 50 mM Tris pH 7.4, 30 mM KCI
2. 10% SDS in Buffer I
3. 10 mM Tris pH 7.4
4. High salt buffer

   0.5 M NaCI, 10 mM Tris, 0.1% SDS pH 7.4
5. Medium salt buffer (wash buffer)

   0.4 M NaCl, 10 mM Tris pH 7.4, 0.1% SDS
6. Low salt buffer 10 mM Tris, 1 mM EDTA pH 7.4, 0.1% SDS
7. 0.1M NaOH
8. Oligo (dT)- Cellulose
9. 2M Sodium acetate pH 5.5

Protocol

Preparation of Oligo - (dT) cellulose column

1. Fill column 3/4th full with Medium salt buffer (buffer 5). Add 0.4M NaCI buffer to 1 g of Oligo- (dT) cellulose in the bottle and shake with Pasteur pipette. Transfer the resin to the column, open the outlet and let buffer drop slowly. Pack in 2 hours. (Use 1g resin, Column should be atleast 5 cm).

2. Remove polysomal pellets from -70^{0}C cabinet. Melt quickly in hands and place in ice bucket. Disperse pellets (6) in 4.25ml of buffer 1. (Add 500 ul of the buffer to each tube and mix with a glass rod gently and pool. Wash the tubes with another aliquot of buffer).
3. Measure O.D. (5 µl of polysome + 995 µl H_2O) at 260 nm. Use the conversion factor 1mg = 10 OD e.g. peak at 260 = 0.48.

   Ribosome concn 48 µg/5ul = 9.6 ug/ul = 9.6 mg/ml or 40.8 mg in 4.25 ml, Polysomes = 40.8 mg.
4. Add 0.25 ml of soln II. and 0.5 ml proteinase K (2 mg/0.5 ml in the same buffer). Cover with parafilm and incubate at room temperature for 2 hours.
5. Degas buffers.
6. Regeneration of oligo -dT column.
   a) Maintain the oligo- dT column head 2cm from top of the column.
   b) Wash the column with 0.1 N NaoH - 15 min.
   c) Wash with sterile water - 15 min.
   d) Pass low salt buffer 10 mM Tris etc. pH 7.4 for 15 min.
   e) Pass medium salt buffer 0.4 M NaCl etc. pH 7.4 for 15 min.
7. Take proteinase K treated polysome preparation in a 50 ml cylinder and make upto 10 ml. with 10 mM Tris (pH 7.4) then upto 50 ml with high salt buffer (0.5M NaCI etc. pH 7.4).
8. Load sample on to the column using a pump. Collect the effluent. It will contain poly $(A)^+$ RNA. Add 0.1 volume of 2M Sodium acetate pH 5.6 and 3 ml of ethanol to precipitate poly (A-) RNA. Leave at -20^{0}C overnight.
9. Wash the column with 100 ml. 0.4 M NaCI pH 7.4.
10. Lower head to within 5 mm of top of column. Elute poly (A^+) with 10 mM Tris-Cl. pH 7.6. Collect in 50 ml cylinder (about 10 ml).
11. Heat at 65^{0}C for 10 min. in order to disrupt the aggregation and immediately plunge into ice bucket. Regenerate column. Repeat steps 8 to 10. Add (high salt 0.5 M NaCI). Buffer to make upto 50 ml.
12. Finally add 0.1 volumes of 2M Sodium acetate pH 5.6 to the 10ml of the poly (A^+) eluant in a siliconized conical flask and add 3 volumes of cold ethanol and store at -20^{0}C overnight.
13. Spin down poly (A^+) and poly (A^-) using Corex tubes in RC 5B at 12,000 rpm at 0^{0}C for 30 min. Wash 2 times with cold 80% ethanol, and dissolve pellet in 50 µl sterile water. Use 1 µl estimate RNA concentration spectro-photometerically.

    1 mg/ml RNA = 25 OD at 260 nm. Freeze rest of the sample in aliquots in liquid nitrogen.

iii) Electrophoresis of ribonucleic acids

RNA mini Gel

This gel system is designed to give a relatively quick estimation of the integrity of the rRNA's from RNA prep. Because of the low buffering capacity of the running buffer, the buffer must be periodically recirculated. The iodoacetic acid is present to inhibit any RNase's that might be around. Because this is not a denaturing gel, the RNA may be stained with ethidium bromide. Load 2 μg, 1 μg, and 0.5 μg of RNA to visualize. This is also a good gel system to look at the flow-through and bound fractions of an oligo-dT cellulose column.

Reagents

- 10X Running Buffer:
- 0.4 M Tris-base
- 50 mM NaOAC
- 10 mM EDTA
- Adjust pH to 7.8 with acetic acid

Protocol

1. Melt 1% agarose in 1X running buffer. When cool (60^{0}C) add 186 mg iodoacetic acid /100 ml of agar.
2. When gel is solid, submerge in 1X running buffer.
3. Prepare RNA for loading by adding 5X loading buffer.

5 1X running buffer

50% glycerol

0.1% bromo-phenol blue

4. Load mini-gel and run at about 75-100 volts with recirculation till BPB is at least 3/4 of the way down the gel. Stain gel in 1 μg/ml ethidium bromide for about 15 min. Rinse gel in distilled water and soak a while if you need to destain.

   The banding pattern of the RNA depends upon the tissue, developmental stage, etc. Generally, the two rRNA bands (28S and 18S) and a smear of mRNA's is visualized. Since the 28S rRNA is approximately twice the size of the 18S rRNA and the two species are present in equal molar amounts, the intensity of the 28S band should be twice that of the 18S.

Note: Soak the gel casting tray, comb, buffer tray in proteinase - K 25 mg/100 ml and 0.10% SDS in sterile water overnight. Rinse them properly and use the pouring gel.

iv) Extraction and electrophoresis of ribosomal proteins

The purification of ribosomal proteins is necessary for an analysis of the structure and function of eukaryotic ribosome. Knowledge of the number and properties of these proteins is also essential for an understanding of the specificity of interaction between ribonucleic acid (RNA) and protein in the ribosome, and of the total function performed by the ribosome in protein biosynthesis. A procedure for the preparation of individual ribosomal proteins by polyacrylamide gel electrophoresis is described here.

Protocol

a) Extraction of ribosomal proteins

1. Resuspend a ribosomal pellet in 1.1 ml of resuspension buffer (10 mM Tris-Cl, pH 7.8, 5 mM $MgCl_2$, 1 mM DTT)
2. Add 5 ml of 100 mM $MgCl_2$ and 3.3 ml of 60% acetic acid. Stir the contents for 45 min at 4^0C
3. Centrifuge at 16,000 rpm for 30 min.
4. Remove the supermatent and dialyse and lyophilize the supernatant.
5. Dissolve the ribosomal proteins in the electrophoresis buffer.

b) Polyacrylamide gel electrophoresis

1. Polymerise 8% polyacrylamide gel containing 6 M urea with following gel solution in a slab gel apparatus. For 100 ml get solution.

Acrylamide	=	8 g
Bisacrylamide	=	0.266 g
EDTA	=	0.800 g
Boric Acid	=	3.2 g
Tris (pH 8.6)	=	4.856 g
Urea	=	36 g
TEMED	=	0.3 ml
APS	=	0.07 g

   Water to 100 ml

2. Pre-run the gel at 10 mA for one hours towards cathode using the following electrophoresis buffer: (1 litre)

Urea	=	6M (350 g)
Borate	=	0.15M (9.3g)
EDTA	=	6.5mM (2.148g)
Tris	=	0.12M pH 8.6 (14.5g)

3. Load sample (200 g protein) using 0.5% methyl green as marker dye.
4. Run at 40 mA per plate or at 10 mA overnight
5. Stain the gel with 0.1% Coomassie brilliant blue R250 in 40% Methanol and 10% acetic acid for half an hour.
6. Destain by shaking with 10% acetic acid and 20% methanol overnight. Observe bands and photograph the gel.

v) Protein estmation

The soluble protein estimation by Lowry et al. (1951) method using Folin reaction has been the most widely used method to estimate the amount of protein (already in solution or solube in dilute alkali) in biological sample. This method is concerned with modifying the Folin phenol reagent procedure by treating protein solutions "with copper in alkali." First of all the protein is pretreated with copper ion in alkali solution and then the aromatic amino acids in the protein of the treated sample reduce the phosphomolybdatephosphotungstic acid present in the Folin reagent. The end product of the reaction has a blue colour. The amount of the proteins in the sample can be estimated via reading absorbance at 750 nm Spectrophotometrically of the end product of the Folin reaction against standard curve of selected standard protein solutions The measurement of protein with copper and the Folin reagent is more sensitive.

Reagents

a) 2 per cent Na_2CO_3 in 0.10 N NaOH.

b) 0.5 per cent $CuSO_4.5H_2O$ in 1 per cent sodium or potassium tartrate.

c) Alkaline copper solution. Mix 49 ml of Reagent A with 1 ml. of Reagent B.

d) Diluted Folin-Ciocalteu phenol reagent, dilute the Folin reagent (about 2-fold) to make it 1 N in acid.

e) Protein Standard Solution (1 mg/ml): Prepare from crystalline bovine serum albumin protein of serial dilutions.

Protocol

1. To a sample of 5 to 100 µl of protein in 0.2 ml. or less in a 5 to 10 ml test-tube.
1. Add 1 ml of Reagent C. Mix well and allow standing for 10 minutes or longer at room temperature.
3. Add 0.10 ml of Reagent D rapidly and mixed within a second or two. Allow to stand for 30 minutes.
4. After 30 minutes or longer, the sample is read in a Spectrophotometer at 750 nm.
5. Calculate the protein concentration from a standard curve prepared from standard protein solutions, if necessary; make appropriate correction for differences between the color value of the working standard and the particular proteins being measured.

Suggested reading

Koundal K.R. (1972). Studies of soluble proteins and isoenzymes during germination in rice (oryza sativa). M.Sc Thesis, Division of Biochemistry, Indian Agricultural Research Institute, New Delhi- 12, India.

Koundal K.R. (1977). Ribosomal proteins of gateway barley and its virescens mutant. Ph.D. thesis, Department of Plant Science, University of Alberta, Edmonton, Alberta, Canada.

Koundal K.R. and Zalik S. (1982). Comparison of cytoplasm ribosomal proteins of Gateway barley and its mutant. Phytochemistry 21: 339–343.

Koundal K.R., Cohen Abe S. and Zalik S. (1980). Reassociation of cytoplasmic ribosomal subunits of barley and its virescens mutant to form active hybrids. Physiol. Plant 50: 194–198.

Lowry H., Rosebrough N., Farr A. and Randall R. (1951). Protein measurement with Folin Reagnt. J. Biol. Chem. 195: 265-275.

8

Preparation of a Probe Radioactive Labeling of Nucleic Acids

DNA fragment to be used as a probe should be purified by gel electrophoresis. The band can be located after staining with ethidium bromide and electroeluted.

The *in vitro* radioactive labeling of DNA fragment is accomplished by the incorporation of a radioactive nucleotide during strand synthesis with Klenow enzyme (large subunit of DNA polymerase-I). The DNA is linearized to which synthetic random hexanucleotides ('primers') added. The Klenow enzyme synhesizews a complementary strand which is radioactivily labeled due to the incorporation of (a-^{32}P) dCTP along with cold ATP, dTTP and dGTP. The radioactively labeled DNA is separated from unicorporated radiolabel dCTP by gel filteration on a Sephadex G-100 column while small molecules like a-^{32}P dCTP diffuse into gel matrix while macromolecules like DNA run very quickly through the column and thus can be separated easily. The probe DNA is precipitated with ethanol and dissolved in 20 μl of TE buffer. A microlitre of this can be again analyzed by a 1% agarose gel electrophoresis together with the dilutions of known amounts of plasmid DNA. By comparing the intensity of DNA bands under UV after ethidium bromide staining, the approximate the approximate concentration of probe fragment can be assessed.

i) Hexanucleotide primed labeling

Reagents

Oligo labeling buffer

To make 0.2 ml of oligo labeling buffer for use with ^{32}P dATP

2M Tris pH 7.5	=	11 μl
1M $MgCl_2$	=	5 μl
0.5M ME	=	10 μl
20mM dCTP	=	5 μl
20mM dGTP	=	5 μl

20 mM TTP	=	5 μl
250 μg hexamer in $T_{10}E_{0.5}$	=	50 μl
St. H_2O	=	109 μl

Protocol

1. Labeling is done in a reaction volume of 50 μl.
2. In an Eppendorf tube take DNA 1 μl (50-100 ng).
3. Add St H_2O to the DNA solution to make upto 29 μl.
4. Boil for 5 min plunge the tube in ice for 1 min.
5. Add this to another tube having the following:

Labeling buffer	11 μl
BSA (5mg/ml)	4 μl
^{32}P dATP	5 μl
Denatured DNA with H_2O from step 4	29 μl
Then add Klenow enzyme	1 μl (20U)
Total vol.	50 μl

6. Incubate at RT for 3 hrs. Store at -20°C for later use.
7. Removal of unincorporated label (optional): Add 1 volume of 4M ammonium acetate pH 4.5 to the labeled DNA mix and vortex. Add 2 volumes of ethanol, mix, chill in ice for 15 min and heat at 37°C for 2 min to redissolve free deoxyribonucleotides precipitated with occasional mixing. Centrifuge at 12,000 x g for 15 min. Wash the pellet once in 90% ethanol and dry the pellet. Redissolve the labeled DNA in TE and use the probe for hybridization.
8. Determination of per cent incorporation (DE 81 filter binding assay): Dilute 1 μl of the labeling reaction with 100 μl of 0.2M EDTA (1:100). Spot 3 μl in duplicate of the diluted sample on Whatman DE 81 2.3 cm circular filters. Dry the filters under a heat lamp. Keep one filter aside for the determination of total cpm in the sample. Then, wash the filter in 50 ml of 0.5 M sodium phosphate, pH 6.8 twice for 5 min. to remove unincorporated label. Dry the washed filter under a hot lamp. Finally, add scintillation fluid to each filter and count in a liquid scintillation counter.

ii) Nick translation

It is a procedure by which radioactive nucleotide phosphates are incorporated into unlabeled DNA for the purpose of making a probe. The reaction depends on the ability of the enzyme DNA polymerase I to initiate DNA synthesis at 3'OH groups which are exposed as nicks in the unlabelled DNA generated in random locations by a limited digest of the DNA with DNase-I. The DNA polymerase synthesizes new DNA in a 5' to 3' direction using labelled triphosphates. The newly polymerized DNA is, therefore, radioactive and can be used in hybridization to detect the presence of homologous sequences.

Requirements

1. Dnase 1: 1 mg/ml in 10 mM HCI, aliquoted and stored at -20^{0}C.
2. DNAse 1 activation buffer: 10 mM Tris. pH 7.6; 5 mM $MgCl_2$ 1 mg/ml BSA
3. 10 x NT Buffer (0.5M Tris, pH 7.8; 5 mM $MgCI_2$; 100 mM Beta mercaptoethanol; 500 µg/ml
4. a-^{32}Pd CTP (>400 Ci/mMole)
5. DNA polymerase I
6. 1mM dATP; 1 mM dGTP, 1 mM TTP, 1 mM dCTP in water, pH 8.0
7. Column buffer (TE buffer including 100 mM NaCl, pH 8.0)

Protocol

- To activate DNAse 1 take 2 µl DNAse 1. Keep on ice 2h.
- Add 18 µl Activation buffer. Immediately before use dilute DNAse 1
- 1:100 in water (5 µl in 495 µl water) then 1:100 dilution again (final dilution 1:10,000)

Prepare Nick translation reaction mix

12 µl	^{32}P dCTP (2 x 10^8 CPM)
2 µl	DNA fragment (0.1 µl)
2 µl	10 x NT buffer
2 µl	dATP, dGTP, dCTP, dTTP mix (18 µM each)
1 µl	DNase 1 40 Pg/ µl
1 µl	Pol 1 1 unit/ µl
20 µl	Total volume

Incubate at 15^{0}C for one hr and stop by adding 1 µl EDTA (0.5M, pH 8.0) and 80 µl of H_2O.

Make up the volume to 100 µl using column buffer. Purify the labeled fragment by passing through a Sephadex G-50 column equilibrated with TE pH 8.0 with

100 mM NaCl. Follow the column by monitoring for radioactivity. Collect radioactive fractions (0.5 ml) into different Eppendorfs. If necessary concentrate by ethanol precipitation or freeze drying. Use 104 to 106 cpm per filter for hybridisation. Denature the pooled DNA in a boiling water bath for 5 min before using it for hybridization.

iii) End labeling

The enzyme T4 polynucleotide kinase specifically transfers gamma ^{32}P-label from ATP to a 5' OH of DNA or RNA. The DNA/RNA species to be labelled must have a 5' OH group. DNA restriction fragments and RNA have 5' phosphate group and thus require dephosphorylation with a phosphatase enzyme before labelling reaction can be carried out. Additionally, a method for labellina on end using terminal deoxynuclotidyl transferase is also possible.

Requirements

- 10 x Kinase buffer: 500 mM Tris pH 7.4, 100 mci $MgCl_2$;
- 3 mM DTT, 10 mM spermidine
- T4 polynucleotide kinase (19 UI/µl)
- gamma ^{32}P-ATP (3,000 µ Ci/mmol;10 µCi/ml)
- 10% SDS

Protocol

1. In an Eppendorftube add the following
    - 1 ml oligonucleotide or DNA fragment (0.5 µg)
    - 2.5 ml 10 x kinase buffer
    - 1.5 µl T4 polynucleotide kinase
2. Behind plexiglass, add gmma ^{32}P-ATP (200 µCi) to the above tube and spin for 15 sec.
3. Incubate at 37^{0}C for 60 min and then stop reaction by heating the tube for 5 min at 65^{0}C.
4. Purify the probe by passing through Sephadex G50 column as described earlier (nick translation).
5. Count 5 µl from different fractions collected. Pool fractions in the leading peak with high counts. Calculate the cpm in the pooled sample and use for hybridization.

Note: Alternatively, the probe can be labeled on 3' end with terminal deoxynucleotidyl transferase. Add 5 µl of 5 x tailing buffer (5mM $CaCl_2$, 1M potassium cacodylate; 100mM Tris-Cl, 5 mM DTT; pH 7.5) and I µg synthetic nucleotide probe in H_2O. After adding 20 µl of Tdt; incubate for 60-100 min and continue as described in step 3 in the above protocol.

Suggested reading

Maniatis T. Fritsch E.F. and Sambrook J. (1988). Molecular Cloning-Laboratory Manual. Cold Spring Harbor Laboratory, New York.

Rigby P.W., Diekmann M., Rhodes C. and Berg P. (1977). Labelling deoxyribonucleic acid to high specific activity *in vitro* by nick translation with DNA polymerase. J. Mol. Biol. 113(1): 237-51.

9

Southern and Northern Blotting and Hybridization

Nucleic acid hybridization requires denaturation i.e. the breakage of H-bridges between the complementary strands of the DNA either by heating at high temperatures or by treatment with denaturing chemicals like formamide. The restricted DNA, therefore, is denatured before transfer from the gel onto synthetic membrane since only single stranded DNA binds to nitrocellulose. Furthermore, the weak interaction between the DNA and the membrane achieved during transfer has is modified into covalent bond by heating to 80^{0}C under vacuum.

A. Southern Blotting and Hybridization

i) Restriction of DNA

Protocol

1. Mix the following in a sterile 1.5 ml Eppendorf tube

DNA	5 µg
10 x buffer	5.0 µl
Restriction enzyme	2.5 µl (10 units / µl)
Sterile H_2O to	50 µl

2. Incubate the mix 37^{0}C for 3 hr.
3. Add 5 µl of 10 x dye to the sample, centrifuge for 5 sec in an Eppendorf centrifuge tube and load on an agarose gel.

ii) Aggarose gel electrophoresis

Reagents

10 x TBE buffer (1 litre)

Tris base	108 g
Boric acid	55 g
EDTA (pH 8.0)	20 ml

Dissolve in DDW and make upto 1 litre.Autoclave before use.

Gel loading buffer (per 10 ml)

Glycerol	5 ml
10 x TBE	1 ml
Bromophenol blue saturated	1 ml
Xylene cyanol 10%	1 ml
H_2O	2 ml

Mix well, divide into 1ml aliquots and store at -20^{0}C.

Protocol

1. Weigh 0.8 g agarose and put it in a 250 ml conical flask. Add 100 ml of 1 x TBE buffer and gently boil the solution in a microwave oven with occasional mixing until all agarose particles are completely dissolved, allow it to cool at 50^{0}C. Prepare the gel mould and keep the comb in position, Pour the cooled gel solution into the gel mould and allow the gel to set for 20 min.
2. Fill the horizontal elelctrophoresis chamber with 1 x TBE. Remove the comb from the gel and place gel with the tray in the electrophoresis chamber.
3. Load the digested DNA sample carefully into the wells. In one well load a standard marker i.e. lambda DNA restricted with Hind III.
4. Run the gel at 20 mA overnight.
5. Stain the gel in 100 ml sterile distilled water containing 1 µg/ml ethidium bromide for 10 min. briefly destain with sterile water.
6. Visualise the DNA bands on a UV transilluminator and place a ruler next to the gel to be able to determine the fragment sizes later on. Take a picture using Polaroid camera with a red filter.

iii) Transfer of DNA to nylon membranes

Protocol

1. After staining and photography, dip gel in 200 ml of 0.25 M HCI for 15 min at room temperature in a glass baking dish.
2. Decant the HCI and denature the DNA by soaking the gel in several vol. of 1.5M NaCI and 0.5 M NaOH for 1hr at room temperature with constant shaking.
3. Decant the denaturing solution and neutralise the gel by soaking in several vol of 1M Tris-CI (pH 8.0) and 1.5M NaCI for 1 hr at room temperature with constant shaking.

4. Wrap a piece of Whatman 3MM paper around a piece of glass plate and place it ninside a large baking dish. Fill the dish with 20 x SSC almost to the top of glass plate and smooth all air bubbles in the 3MM paper with a glass rod.
5. Invert the gel so that its original underside is now upper most and' place it on the wet 3MM paper. Remove air bubbles if any between the gel and the paper.
6. Cut a piece of nylon membrane about 1-2 mm larger than the gel. Dip it in sterile water and then float the membrane in 20 x SSC for 5 min.
7. Place the wet nylon membrane on top of the gel and remove any air bubbles that are trapped between the gel and the membrane.
8. Wet two pieces of Whatman 3MM paper, cut to exactly the same size as the gel in 20 x SSC and place it on top of the nylon membrane followed by 6 cm stack of dry paper towels. Put a glass plate on top of the stack and weight it down with 500 g and 3 MM paper under the gel, surround the gel with a water tight border of Saran wrap.

   Note: Do not touch nylone membrane with naked hands. Always wear gloves and use forceps while handling nylon membranes.
9. Allow transfer of DNA to proceed for 7 hr to overnight. Small fragments of DNA (< 1 kb) transfer from 0.8% gel within an hour or two while transfer of DNA > 15 kb takes 15 h or more.
10. Disassemble the blot in reverse order and using a soft pencil clearly label the slots on the membrane.
11. Soak the membrane in 6 x SSC at room temperature for 5 min.
12. Place the filter on 3MM paper to air dry.
13. Bake the membrane at 80^{0}C for 1 h in a vacuum oven.
14. Soak the membrane in warm 2 x SSC, seal it in plastic bag and store in dark in refrigerator until use.

iv) Prehybridization

Reagents

1. 20 x SSC: 3M NaCI, 0.3M Na citrate pH 7.0.
2. Non homologous DNA, 2 mg/ml Salmon sperm DNA. Prepare by dissolving in water overnight, then sonicate. Store at 4^{0}C.
3. 100 x Denhardts': 2% (w/v) Ficoll (400,000).
4. 10% SDS.

5. 1M Na phosphate buffer, pH 6.5.
6. Formamide deionized using a mixed bed ion exchange resin such as Bio- Rad AG 5001-X8.

Protocol

i) Prepare hybridisation solution which contains 5 x Denhardts', 5 x SSC, 50 mM phosphate pH 6.5, 0.1% SDS, 250 pg/ml non-homologous DNA and 50% formamide. Incubate the solution for 10 min at 95-100^{0}C to denature DNA, then cool on ice.

ii) Keep the nylon membrane in a plastic bag and pour hybridisation solution (4 ml/100 cm^2) and seal the bag. Seal the bag inside a second bag containing damp paper towel.

iii) Immerse the bags in 42^{0}C water bath for 1-2 hr.

v) Hybridization

Reagents

i) Open the bag containing the membrane and remove the hybridisation solution.

ii) To the membrane in the bag add fresh 2 ml/cm^2 hybridisation solution to which denatured probe has been added (denatured by heating at 95-100^{0}C for 10 min).

iii) Seal the bag and massage in order that the solution is distributed well.

iv) Immerse the bags in water bath at 42^{0}C for 8-10 hr or for the duration of the hybridization.

vi) Washing the membrane and autoradiography

i) After completion of the hybridisation, remove the membrane from the bag and wash briefly in 2 x SSC and 0.1% SDS.

ii) Keep the membrane in a fresh bag and pour 2 x SSC + 0.01% SDS buffer (250 ml/100 cm^2 membrane). Seal the bag and agitate vigorously for 5 min at room temperature.

iii) Repeat Step (ii) three times.

iv) Add 0.1 SSC + 1% SDS @ 250 ml/100 cm^2.

v) Repeat Step (IV) twice more.

vi) Remove the membrane from the bag and keep it in between Saran wrap. Remove any remaining liquid by pressing with tissue paper and expose to X-ray film.

(vii) Autoclave the film before disposal.

Autoradiography

Radioactive nucleic acids and proteins may be detected by autoradiography or by liquid scintillation spectroscopy. Autoradiographic methods however are considered superior as they are more sensitive and give higher resolution.

Requirements

i) X-ray film developer

Water at 52^{0}C	500 ml
Metol	1 g
Sodium Sulphite	75 g
Hydroquinone	9 g
Sodium carbonate (monohydrate)	25 g
Potassium bromide	5 g

Dissolve in the given order and make up to 1 litre with cold water.

ii) Fixer

Sodium thiosulphate	80 g
Potassium metabisulphate	5 g

Dissolve in the given order in 250 ml and filter

iii) Stop bath (water containing 2% acetic acid)

iv) X-ray film

v) Casette with intensifying screen

Readymade developer and fixer are also available.

Protocol

1. Tape the material to be autographed to a backing of thin card board or Whatman 3 MM paper.
2. Place pieces of tape marked with radioactive ink at several locations around the edge of sample.

   (Radioactive ink is made by mixing a small amount of ^{32}P with a water proof blackink. It is convenient to make the ink in three grades:

   Very hot - 2000 cps on a minimonitor

   Hot - 500-2000 cps on a minimonitor

   Cool - 50-500 cps on a minimonitor

3. When the ink is dry, wrap the sample and backing sheet in Saran Wrap. This protects the casette and intensifying screen from contamination with penetrating radioactivity. The plastic wrap may block a significant amount of the ^{35}S signal. If ^{35}S is used as the radioactive source, do not cover with the plastic wrap unless sample is sufficiently radioactive to penetrate plastic wrap.
4. In a dark room, place the sample in a X-ray film casette and cover it with a X-ray film (use the screen if required. The sensitivity of the film may be increased if intensifying screen is used).
5. Close the cassette and wrap it in black bags.
6. Keep at -70^{0}C for several hrs to a few days (this depends on the radioactivity).
7. Remove the film after exposing and develop it in developer for 3-5 min. in dark.
8. Wash the film in stop bath for 1 min and fix the film in fixer for 5 min in dark.
9. Wash under running water for 30 min.
10. Dry the film at room temperature.

vii) Transfer of DNA from phage plaques

1. Grow phage on NZ plates with approximate bacterial cell lawn with 10^2-10^5 plaques. Use dry or 2-day-old plates and top agar having a concentration of 0.9% agar.
2. Cool the plates at 4^{0}C for about an hour to harden the agar.
3. Place a nitrocellulose filter on the cell lawn having phage plaques taking care not to allow air bubbles to form. To do this, slightly curve the filter, placing the center down to the center of the plate. Then slowly lower the edges to the agar.
4. Adsorb for 1-5 min.
5. Mark the filter orientation by stabbing the filter with a syringe needle containing India ink. Nark the bottom of the plate at the corresponding sites.
6. Carefully remove filter from plate.
7. Dip filter or batch wash in denaturing solution for about 5 min. For this, float the filter (plaque side up) on the solution for a couple of min, then submerge it by rocking the solution.
8. Dip filter or batch wash in neutralizing solution for about 5 min.
9. Rinse filter in 2 x SSC.
10. Proceed to hybridization as described earlier.

viii) Transfer of DNA from bacterial colonies

1. Grow approximate bacterial colonies on dry or 2-day-old nutrient agar plates. These may be gridded or may be from the original transformation.
2. Place a dry nitrocellulose filter on top of the colonies while they are quite small.
3. Mark orientation of the filter and plate as with plaque lifts and transfer the filter (colony side up) to a fresh plate. Grow these colonies for a couple of hours or until visible growth has occurred. The original plate will serve as the master plate and should be incubated to allow the colonies to grow.
4. Place the filter, colony side up, on two sheets of Whatman 3MM paper which has been saturated with denaturing solution. Allow to sit about 10 min.
5. Transfer filter to a dry sheet of 3MM paper, colony side up, to remove excess liquid.
6. Place filter, colony side up, on 2 sheets of 3MM paper that has been saturated with neutralization solution. Allow to sit about 10 min. Repeat steps 5 and 6.
7. Rinse filter in 2 x SSC and dry in a vacuum oven at 80^{0}C for 2 hr.
8. Proceed to hybridization as described earlier.

B. Northern Blotting and Hybridization

In order to accurately determine its size, the RNA must be electrophoresed under denaturing conditions. Additionally, denatured samples bind much better to the filter. Three treatments are commonly used; formaldehyde, glyoxal and methyl mercury hydroxide. Methyl mercury is very toxic and should be used only when RNA having biological activity is to be recovered from the gel. Protocols for formaldehyde and glyoxal gels are given below. Formaldehyde gels are preferred.

i) Agarose gel electrophoreses of RNA

Reagents

- 1 x = 0.02% Ficoll
- 0.02% Polyvinylpyrollidone
- 0.2% BSA
- May be made up 100 ml with H_2O

Protocol

Sample preparation and agarose gel electrophoresis

Formaldehyde method: Heat RNA in 10 mN $NaPO_4$ buffer (pH 7.4), 2.2 M formaldehyde, 50% formamide, 0.5 mM EDTA at 65°C for 5 min. Cool to room temperature and add 5 x loading buffer (0.5% SDS, 25% glycerol, 25 mM DT, and bromophenyl blue). Electrophorese on 0.8-1.5% agarose gel poured in 1.1M formaldehyde, 10 mM $NaPO_4$ (pH 7.4), at room tempreture.

The gels are run at about 60 V with buffer recirculation.

Formaldehyde is obtained as a 37% solution in water (12.3 M). The pH should be greater than 4.0.

Glyoxal method: Incubate RNA in 1.0M glyoxyl, 10 mM $NaPO_4$ (pH 6.5) for 1 hr at 50°C. Add 1/10 vol of 10 x loading buffer (0.3% bromophenyl blue, 50% glycerol). Fractionate through a 0.8-1.5% agarose gel in 10 mM $NaPO_4$ at 50 V, at room tempreture.

Glyoxal is purchased as a 7M solution. It should be deionized with Amberlite MB-3 resin. Place about 10 ml of resin in a 50 ml tube and add 40 ml of glyoxyl. Shake and let resin settle. Repeat several times. Transfer glyoxal to a second tube and add fresh resin. Repeat transfers till resin no longer exhibits colour change. Aliquot glyoxal to small tubes and freeze at -70°C.

ii) Transfer of RNA

Protocol

1. Rinse the RNA gel after electrophoresis for 20 min each in two changes of 500 ml of 10 x SSC to remove formaldehyde from gel.
2. Place a 20 x 30 cm glass plate over tray. Drape wetted filter over plate with both ends hanging into buffer to act as wick as described for Southern blotting.
3. Carefully place gel upside down over wick.
4. Wet nitrocellulose (NC) filter in place on top of gel, smooth to remove air bubbles.
5. Wet two pieces of Whatman 3 MM paper in H_2O, place individual filters flat on top of NC filter and smooth to remove air bubbles.
6. Place six paper towels flat over 3 MM Whatman paper. Stack four piles of folded paper towels, each 2-3 cm thick.
7. Cover with 20 x 30 cm glass plate. Place a small weight on top. Allow transfer to nitrocellulose by capillary action to proceed for 16-18 hr at room temperature.

8. Next day, remove paper towel stacks and paper sheets. Mark first well with two pinholes to determine the filter orientation.
9. Remove gel and discard. Place NC filter between two sheets of Whatman 3 NM filter. Dry it for 2 hr and ready for hybridization.

iii) Prehybridization, hybridization and autoradiography

Protocol

1. Place filters in a seal bag.
2. Pre-hybridize filters in hybridization solution. Squeeze out bubbles and seal. Incubate at 65°C (or 4°C) overnight.
3. Cut open bag and squeeze out hybridization solution.
4. Boil sufficient probe to give a final concentration of 10^5-10^6 cpm/ml hybridization solution (5 min.).
5. Add probe to hybridization solution and mix well.
6. Add to bag, squeeze out air bullies and seal. Incubate 65°C (or 42°C) for 16 hr.
7. Cut open bag, remove filter and wash in a large vol according to the following regime:
   - 2 times 15 min each 1 x SSC; 0.1% SDS (room temperature).
   - 3 times 30 min each 0.2 x SSC; 0.1% SDS (32-42°C).
8. Check that filters are 'cold' with gieger counter. Air dry briefly, cover with plastic wrap and put up to film for auto radiography described earlier.

Suggested reading

Ausubel F.M., Brent R., Kingsten R.E., Moore D.D., Seedman J.G. Smith, J.A. and Struhi K. (1995). Short Protocols in Molecular Biology, John Wiley and Sons Publishers.

Reed K.C. and Mann D.A. (1985). Rapid transfer of DNA from agarose gels to nylon membranes. Nucleic Acids Res. 13: 7207-21.

Southern E.M. (1975). Detection of specific sequences among DNA fragments separated by gel electrophoresis. Journal of Molecular Biology 98: 503-17.

10

Construction of Genomic Library

Exchange of genetic material rarely occurs between unreleated organisms and even similar species may not exchange chromosomal genes. A certain degree of DNA sequence homology is a prerequisite for genetic recombination in *vivo*. However, recombination between DNA molecules *in vitro* is not subject to such taxonomic restrictions. Consequently using modern methodology of recombinant DNA technology, genetic material from organisms of quite diverse origins with little DNA sequence homology can now be introduced into a host cell for propagation and expression through appropriate vectors. In this way, a gene of interest can now be isolated from any living source or synthesized *in vitro*. The isolation of a gene begins with the cloning of fragmented DNA of the organism into appropriate vector. The set clones representing the whole genome of an organism is called genomic library. The genomic library is screened by nucleotide probe or antibody to find and isolate the appropriate gene.

Solutions

1. 10 x *Sau 3A* buffer

1 M Tris pH 7.5	60 µl
0.2M $MgCl_2$	300 µl
5.0M NaCI	100 µl
Water to	1 ml

2. Sucrose 10% solution

Sucrose	10 g
2M Tris pH 7.4	1.0 ml
100 mM EDTA	10.0 ml
5N NaCI	20.0 ml
Water to	100 ml

3. Sucrose 40% solution

Sucrose	40 g

2M Tris pH 7.4	10.0 ml
100 mM EDTA	10.0 ml
5N NaCl	20.0 ml
Water to	100 ml

4. Phenol redistilled and equilibrated.

   To 500 ml phenol (redistilled), add 100 ml 2M Tris pH 7.4 and 130 ml sterile water. Mix well. Allow the two layers to separate. Remove aqueous upper phase. Then add:

2M Tris pH 7.4	100 ml
m-Cresol	25 ml
β-mercaptoethanol	1 ml
8-Hydroxyquinoline	500 mg

   Store aqueous and phenol layers together in a hood at room temperature.

5. EMBL-3 Cloning Kit
6. 10 x Ligation buffer

Tris pH 8.0	0.5 M
$MgCl_2$	70 mM
DTT	10 mM

7. ATP 10 mM
8. NZY Broth (litre)

NaCl	5 g
$MgSO_4$, $7H_2O$	2 g
Yeast extract	2 g
NZ amine (casein hydrolysate) Autoclave	10 g

9. Top agar

   Same as NZY broth except with 0.7% agarose. Autoclave.

10. SM buffer (per litre)

NaCl	5.8 g
$MgSO_4$, $7H_2O$	2.0 g
1M Tris pH 7.5	50.0 ml
2% gelatin	5 ml

   Autoclave

11. NZY plates

    Add 15 g Difco agar/litre of NZY broth, autoclave and pour in plates.

12. TB-broth (per litre)

NaCl	5.0 g
Bactotryptone	1.0 g
Autoclave	

13. 20% Maltose

Maltose	20 g
Water	100 g
Filter sterile	

Protocol

i) Partial digestion of plant DNA with *Sau 3A*

Isolate genomic DNA as described earlier.

An analyticl test reaction is first carried out to determine the conditions for generating DNA fragments of around 20 kb sizes. This is done as follows:

1. Prepare reaction mixture by adding the following:

Eukaryotic DNA	25 µl (10 µg)
dH_2O	125 µl
Final volume	150 µl

2. Dispense 30 µl into an Eppendorf tube (Tube-1). Dispense 15 µl each into tubes 2-8. Dispense the remainder into Tube-9. Chill all tubes on ice.
3. Add 4 units of Sau3A to Tube-1 and mix well (2 units/pg DNA). Transfer 15 µl of the reaction mix from Tube-1 to Tube-2, mix well and continue two fold serial dilution through to tube-8. Tube-9 will have unrestricted DNA.
4. Incubate tubes 1-8 at 37^{0}C for 1 h.
5. Stop reaction by adding 1 µl of 500 mM EDTA and chilling to 0^{0}C. Add 1.5 µl gel loading dye to each sample and load on 0.4% agarose gel. Electrophore at 40V overnight along with suitable size marker.
6. Photograph the gel and compare with the markers. Determine the amount of enzyme needed to produce the maximum intensity of fluorescence in the 15-20 kb region on the gel and use that concentration for large scale preparation of partially digested DNA.

ii) Large scale preparation of partially digested DNA

1. Scale up the optional digestion conditions to digest 1 mg of genomic DNA

   In a 10 ml tube, add

1 mg genomic DNA	1 ml
10 x *Sau3A* buffer	250 μl
Sau 3A	2 units / μg DNA
Water to	2.5 ml

2. Incubate at 37^{0}C for 1 hr.
3. Extract DNA with 2.5 ml of 1:1 mixture of phenol and chloroform. Remove aqueous phase after centrifugation.
4. Add 7 ml ethanol to aqueous phase. Precipitate on ice for 10 min.
5. Centrifuge at 12,000 x g for 10 min at 4^{0}C.
6. Decant ethanol. Resuspend DNA pellet in 500 μl of TE buffer. Leave at 4^{0}C for DNA to dissolve properly.

iii) Size fractionation by sucrose density gradient

1. Prepare a 38 ml, 10-40% linear sucrose density gradient in a Beckman SW-27 polyallomer tube (or its equivalent).
2. Heat DNA sample for 10 min at 68^{0}C. Cool to 20^{0}C and load on to the gradient.
3. Centrifuge at 26,000 rpm for 24 h at 20^{0}C.
4. Puncture bottom of the tube and collect 0.5 ml fractions.
5. Mix 10 μl of every third fraction with 10 μl of H_2O and 2 μl gel loading dye and analyse on 0.4% agarose gel using proper size markers. Adjust the sucrose and salt concentrations of the markers to correspond that of sample.
6. After electrophoresis, pool the gradient containing DNA fragments in the 15-20 kb size range and dialyse against TE buffer at 4^{0}C for 12-16 h with one buffer change after at least 4-6 h. Leave enough space in the bag for the sample to expand two to three fold.
7. Concentrate the dialysed DNA by extraction with equal vol of 2-butanol until the vol is reduced to 5 ml. Precipitate DNA with ethanol and leave at -20^{0}C for one h.
8. Pellet the DNA and dissolve in TE and analyse an aliquot on 0.4% agarose gel.

iv) Cloning of DNA to EMBL-3 arms

1. Ligation

   Add the following:

   i) 1.0 µl (1 µg) EMBL 3 predigested arms

   ii) 0.8 µl (0.4 µg) of DNA insert

   iii) 0.5 µl 10 x ligase buffer

   iv) 0.5 µl 10 mM ATP

   v) 0.5 µl (3 Weiss units) T-4 DNA ligase

   vi) H_2O upto 5.0 µl volume

2. *In vitro* packaging

   i) Quickly thaw the extracts 1 yellow tube (sonic extract), 1 red tube, (freeze thaw lysate) provided in the kit on ice.

   ii) Add DNA (0.2 to 5 µg) to red tube.

   iii) Add 15 µl from yellow tube into red tube containing DNA.

   iv) Gently mix the contents

   v) Incubate at room temperature for 2 h.

   vi) Add 500 µl SM buffer and 20 µl chloroform. Store at 4°C

3. Plating

   i) Inoculate single colony of LE 392 and P2 392 strains and grow overnight with vigorous shaking at 30°C in TB broth supplemented with 0.2% Maltose and 10 mM $MgSO_{4.}$

   ii) Spin the cells at 1000 x g for 10 min and gently resuspend in 0.4 vol of 10 mM $MgSO_4$

   iii) Adjust the OD_{600} of the cells to 0.5.

   iv) Make serial dilutions of the packaging extract with SM buffer.

   v) Plate aliquots of the packaged ligation product on host strains LE 392 and P2 392. Only recombinants will grow on P2 392 cells. Mix 200 µl of cell (OD_{600} 0.5) with appropriate volumes of diluted packaged product. Allow the phage particles to absorb to the bacteria by incubating at 37°C for 15 min. Then add 3 ml of top agarose to solidify and incubate at 370C. Plaques should be visible after incubation for 8 to 12 h at 37°C.

4. Amplification of Library

i) Mix aliquots of the packaged lambda clones containing approximately 50,000 recombinant phages with 600 ul of host cells (prepared as described above at 3(i).

ii) Incubate tubes for 15 min at 37⁰C.

iii) Add 6.5 ml of top agar (48⁰C) to each tube of infected cells and spread evenly onto NZY 150 mM petri dishes of bottom agar.

iv) Incubate plates at 37⁰C for 5-8 h. Do not allow the plaques to grow too large.

v) Pipette 10 ml of SM buffer onto each plate and incubate at 4⁰C overnight with gentle shaking.

vi) Next day recover the bacteriophage suspension frem each plate and pool into a sterile glass container. Rinse the plates with 2 ml of SM buffer and pool. Add chloroform to 5%. Incubate for 15 min at room temperature.

vii) Centrifuge at 4,000 x g for 5 min to remove cell debris.

viii) Recover the supernatant, pool and transfer to sterile polypropylene or glass bottle. Add chloroform to 0.3% and store at 4⁰C.

ix) Freeze a portion of the amplified library after the addition of 7% (v/v) DMSO. Store at -70⁰C.

v) Titering of library and plating lysate stocks

1. Make a few consecutive 10^{-2} dillutions of the above supernatant.
2. Plate 10 µl of the 10^{-4} dilution in the host strain.
3. Incubate atleast for 8 hrs. at 37⁰C
4. Count the plaques. You should get approximately 400 plaques on the 10^{-4} dilution plate with Giga pack II Gold packaging extract.

   To achieve maximum yields, the number of bacteriophages plated should be so adjusted that the outer edges of the expanding plaques just touch after approximately 12 hrs of incubation. An inoculum of 10^5 plaque forming units (pfu) is usually sufficient to produce confluent lysis on a 85 mm plate (567 cm^2). By the end of the period of growth, there should be no patches of uninfected bacteria.

1. Mix 10^5 pfu of bacteriophage (or 1/20 of a resuspended plaque) with 0.1 ml of plating bacteria. Incubate at 37⁰C for 20 min.
2. Add 3.0 ml of melted top agar/agarose at 47⁰C, mix, and pour onto a labeled 85 mm plate containing 30 ml of hardened bottom LB agar. Freshly poured plates give the best results.

3. Invert and incubate the plate for 8-12 hrs, until lysis is confluent.
4. Turn the plate over, add 5 ml of SM, and store the plate at 4^0C for several hours with intermittent, gentle shaking.
5. With a Pasteur pipette, harvest as much as possible of the SM. Add 1 ml of fresh SM and store the plate for 15 min in a tilted position to allow all the fluid to drain into one area. Again remove the SM and combine it with the first harvest. Discard the plate.
6. Add 0.1 ml of chloroform to the pooled SM, vortex briefly, and centrifuge at 4000 x g for 10 min at 4^0C.
7. Recover the supernatant and add chloroform to 0.3%. The titre of the bacteriophage (app. 10^{10}/ml) usually remains unchanged as long as the stock is stored at 4^0C.

Suggested reading

Koundal K.R. Sharma R.P. and Mehta S.L. (1989). Construction of chickpea genomic library. Indian J. Expt. Biology 27: 858-60.

Old R.W. and Primrose S.B. (1994). Principles of Gene Manipulation. Blackwell Scientific Publication, London.

Sambrook J., Fritsch F.F. and Maniatis T. (1989). Molecular cloning-A laboratory Manual, Cold Spring Harbor Laboratory Press, New York.

11

Screening of Genomic Library

Recombinant DNA is advancement in molecular technology which allows manipulation of genomes of higher organisms and dissection of their complex regulatory circuits. This new methodology is being used to understand the biochemistry of living organisms. The isolation of gene begins with the cloning of fragmented DNA of the organism into appropriate vector. The sum of all the genes as individual clones representing the whole genome of an organism is called genomic library. The process of identifying the particular clone containing the gene of interest from among the very large number of others in the gene library using nucleic acid probe based on hybridization with nucleic acids is known as screening of library to find the appropriate gene.

Reagents

1. 0.5 M NaOH

   1.5 M NaCl
2. 0.5 M Tris-HCl pH 7.4

   1.5 M NaCl
3. 20 X SSC

   NaCl 175.2 g

   Na_3 Citrate 88.4 g

   Double distilled water 1000 ml
4. 100 X Denhardts

   2% BSA Bovine Serum Albumin

   2% Ficoll (Sigma 400.000)

   2% PVP
5. Salmon sperm DNA (10 mg/ml)
6. 500 mm EDTA pH 8.0
7. TE (10 mM Tris-HCl pH 8.0,1 mM EDTA)

8. Hybridization Solution

   6 x SSC

   5X Denhardt solution

   0.1% SDS

   Prehybridization Solution

   6 x SSC

   5X Denhardt solution

   0.1% SDS

   50 μg/ml Salmon sperm DNA in H_2O

Media

LB Medium (per litre)

Bacto-tryptone	10 g
Bacto-yeast extract	5 g
NaCl	10 g

Adjust pH to 7.5 with sodium hydroxide

Maltose

Maltose	20 g
H_2O to	100 ml

Sterilize by filtration through a 0.2 micron filter cassette. Add 1 ml of sterile, 20% maltose solution for every 100 ml of medium

SM (per litre)

NaCl	5.8 g
$MgSO_4$-$7H_2O$	2 g
1 M Tris-HCl (pH 7.5)	50 ml
2% gelatin	5 ml

Sterilize by autoclaving and store in 50 ml lots

Media Containing Agar or Agarose

Just before autoclaving, add (per litre one of the following to appropriate medium

Bacto-agar	14 g (for plates)

Bacto-agar	7 g (for top agar)
Agarose (type-1, low EEO)	15 g (for plates)
Agarose	7g (for top agarose)

Protocol

i) Preparation of plating bacteria

1. Inoculate a single bacterial colony from a day old plate of the indicator bacteria into 50 ml of LB medium supplemented with 0.2% maltose, in a 250-ml flask and grow to an OD_{600} = 0.8. Bacteria grown in the presence of maltose adsorb bacteriophage λ more efficiently than if maltose was not present; the sugar induces the maltose operon, which contains the gene (lambda) coding for the λ receptor.
2. Centrifuge the cells at 4000g for 10 min at room temperature.
3. Discard the supernatant and resuspend the cell pellet in sterile 0.01 M $MgSO_4$ (0.4x the volume of the original culture). The resuspended cells are diluted to an OD_{600} = 2; i.e., 1.5 x 10 cells/ml).
4. Store at 4^0C. The bacterial suspension may be used for up to one week. However, the highest plating efficiencies are obtained with fresh cells (0-2 days old)

ii) Plating of library

1. Prepare 100-fold serial dilutions of bacteriophage stocks in SM. Dispense 0.1 ml of each dilution tc be assayed into each of two 13-mm x 100-mm test tubes.
2. Add 0.1 ml of plating bacteria to each tube. Mix by shaking or vortexing. Incubate at 37^0C for 20 min to allow the bacteriophage particles to adsorb.
3. Add 3.0 ml of medium (47^0C) containing melted 0.7% agarose to the first tube, vortex gently, and immediately pour onto a plate containing 30-35 ml of hardened bottom agar medium. Try to avoid air bubbles. Swirl the plate gently to ensure an even distribution of bacteria and top agarose. Repeat with each of the tubes. Close the plates and let them stand for 5 minutes at room temperature to allow the top agarose to harden. Invert the plates and incubate at 37^0C. Plaques begin to appear after about 8 hours and should be counted or picked after 12-16 hours of incubation.
4. Bacteriophages library usually contains 5 x 10^{10} - 10^{11} phages/ml.

iii) Screening plaques by hybridization

1. Mix aliquots of the library in bacteriophage λ containing 20,000 plaque frming units in a volume of 50 μl or less with 0.3 ml of freshly prepared plating bacteria.

2. Incubate at 37⁰C for 20 min.

3. Add 6.5 ml of molten (48⁰C top agarose (0.7%) and pour onto a 150 mm LB agar plate. The plates must be dry, otherwise the layer of top agarose peels oh with the filter.

4. Incubate at 37⁰C until the plaques reach a diameter of approximately 1.5 mm and are just beginning to make contact with one another (10-12 hrs). The plate should not show confluent lysis.

5. Chill the plates at 4⁰C for two to four hours to allow the top agarose to harden.

iv) Transfer of plaques to flters

1. Wear gloves and then number nitrocellulose filters with a soft pencil.

2. At room temperature, place a dry nitrocellulose circle neatly onto the surface of the top agarose of a cold plate with plaques so that it comes into direct contact with the plaques. Be careful not to trap air bubbles. The filter should be handled with gloved hands; finger oils prevent wetting of the filter and affect transfer of DNA. Mark the filter in three or more asymmetric locations by stabbing through it and into the agar beneath with an 18-gauge needle attached to a syringe containing waterproof black drawing ink.

   Once in contact with the top agarose, the filter wets very rapidly and transfer of bacteriophage occurs quickly. Therefore, do not move the filter, once contact with the plate is made. The easiest way of placing the filter on the plate is to hold it by its edges bending it slightly so that the middle of the filter makes contact with the center of the plate. Let wetting action pull the rest of the filter onto the plate.

   Make certain that the keying marks are asymmetrically placed and that both the filter and the plate are marked. There must be enough ink on the plate to be easily visible when a second filter is in place. Large blotches of ink, however, are undesirable.

3. After 60 seconds, use a blunt-ended forceps to peel off the first filter and immerse it, phage side up, in a shallow tray of a denaturing solution (1.5 M NaCl, 0.5 M NaOH) for 2 min. This process causes denaturation of the protein portions of the phage and releases the DNA which binds to the

nitrocellulose filter. Transfer the filter into neutralizing solution (1.5 M NaCl, 0.5 M Tris- CI, (pH 7.0) for 2 min. Rinse the filter in 4 x SSC (diluted from 20X SSC) for 5 min and place it on Whatman paper to dry with DNA side up for 30 min.

4. Place a second, dry filter onto the same plate and mark it with ink at the same locations. After 1-2 minutes, peel the filter off the plate. Denature the DNA and neutralize as described in step 3 above.
5. After all the filters are dry, wrap them between sheets of Whatman 3MM paper. Fix the DNA to the filter by baking for 2 hours at 80^{0}C in a vacuum oven. (Overbaking can cause the filters to become brittle). Store in a desiccator until ready to use.

v) Hybridization with labelled probe

Prehybridization and hybridization

Seal each filter into a plastic bag. Cut a corner of the sealed bag and add the prehybridization solution to each bag. For a small filter (82 mm) add 5 ml of prehybridization solution per bag, while for a 147 mm of filter add 7 ml of prehybridization solution to each bag. Squeeze all the air bubbles out and then seal the corner of the bag. Each of these bags containing a single filter should be put into a larger bag and sealed in all four corners. Put this big plastic bag containing all the small filters on a platform of G-76 shaker. Put lead weights on all four corners and fill the shaker with 60^{0}C water so that the bag is immersed. Let it shake at a setting of 4.5 for four hrs. After four hours, remove bag from water bath. Split the big bag and take out all the small bags. Cut the corner of each small bag and remove all the prehybridization solution. Add 2 ml of the hybrldization solution per bag. Add ^{32}P-labelled probe. Seal the corners after removing all the bubbles. Put all the small bags into a big bag. Seal all four corners and immerse in a 60^{0}C water bath shaker for 16 hours.

Washing of Filters

1. After 16 hrs of hybridization at 60^{0}C take the filters out of the bags.
2. Wash twice for 30 min at 37^{0}C in 1 litre of 6 x SSC 0.1% SDS.
3. Wash two times in 1 litre of 6 x SSC for 30 min at 60^{0}C.
4. Put the filters on 3 MM Whatman paper to air dry with plaques side up and numbered side down. Let them air dry for one hour.
5. Tape S and S filters on a sheet of 3MM Whatman chromatography paper with DNA side down and markings up. Make four radioactive markings on 3MM Whatman paper for later aligment of the radioautographed film. Wrap

the 3MM paper with S&S filters in Saran Wrap. Load into casettes with X-ray film for autoradiography in a dark room. Put at -80°C for overnight exposure and then develop the film.

vi) Interpretting plaque DNA hybridization signals

1. Put 3MM. filter paper containing hybridized filters on a light box.
2. Put the developed film on the filter paper and align with the radioactive ink marks.
3. Mark on the film the number of each of the filters and also note the ink marks which match on the hybridized filter, the agar plate and the developed film.
4. Put the film on a light box and the agar plate with plaques on top. Align the marks on the agar plate and the film.
5. With Pasteur pipette, pick the plaques which give positive signals.

vii) Picking positive plaques

1. Place 1.0 ml of SM in a 13-mm x 100-mm polypropylene tube. Add 1 drop of chloroform.
2. Using a Pasteur pipette equipped with a rubber bulb, or a micropipette, stab through the chosen plaque into the hard agar beneath. Apply mild suction so that the plaque, together with the underlying agar, is drawn into the pipette.
3. Wash out the fragments of agar into the 1.0 ml of SM containing a drop of chloroform. Let stand at room temperature for an hour or two to allow the bacteriophage particles to diffuse out of the agar. An average plaque yields 10^6-10^7 infectious phage particles, which can be stored indefinitely at 4°C in SM-Chloroform without loss of viability.

Note-Because bacteriophage λ can diffuse considerable distances through the top agar layer, choose well-separated plaques. For the same reason, it is advisable to pick plaques shortly after the bacterial lawn has grown and the bacteriophage plaques have first appeared.

viii) Secondary screening

Plate these plaques at a dilution of 10^{-2}, 10^{-4} and 10^{-6} on LB agar plates. Choose agar plates with well dispersed well isolated plaques. Transfer these plaques to cellulose filters and hybridize against the defined gene. After autoradiography, a well isolated recombinant plaque which hybridizes to the probe is picked horn each plate for further work.

ix) Plating lysate stocks

To achieve maximum yields, the number of bacteriophages plated should be adjusted so that the outer edges of the expanding plaques just touch after approximately 12 hrs of incubation. An inoculum of 10^5 plaque-forming units (pfu) is usually sufficient to produce confluent lysis on a 85-mm plate (567 cm^2). By the end of the period of growth, there should be no patches of uninfected bacteria.

1. Mix 10^5 pfu of bacteriophage (or 1/20 of a resuspended plaque) with 0.1 ml of plating bacteria. Incubate at 37°C for 20 min.
2. Add 3.0 of melted top agar/agarose at 47⁰C, mix, and pour onto a labeled 85-mm plate containing 30 ml of hardened bottom LB agar. Freshly poured plates give the best results.
3. Invert and incubate the plate for 8-12 hrs until lysis is influent.
4. Turn the plate over, Add 5 ml of SM, and store the plate at 4°C for several hours with intermittent, gentle shaking.
5. With a Pasteur pipette, harvest as much as possible of SM. Add 1 ml of fresh SM and store the plate for 15 min in a titled position to allow all the fluid to drain into one area. Again remove the SM and combined it with the first harvest. Discard the plate.
6. Add 0.1 ml of chloroform to the pooled SM, vortex briefly and centrifuge at 4000 x g for 10 min at 4°C
7. Recover the supernatant and add chloroform to 0.3%. The titer of bacteriophage (approximately 10^{10}/ ml) usually remains unchanged as long as the stock is stored at 4°C.

Master stocks of important ë bacteriophages (e q. vectors), whose genotype has been veriied should also be stored at - 70°C. Add DMSO to bacteriophage stock to a final concentration of 7% (v/v). Mix gently. Plunge the container into figured nitrogen. When the liquid has frozen, transfer the continer to a freezer at -70°C for long term storage.

To recover the bacteriophage, scrape the frozen surface of the liquid with a sterile 18 gauge needle, streak the needle, over the surface of a plate containing indicator bacteria in order to obtain bacteriophage plaques.

Suggested reading

Sambrook J., Fritsch F.F. and Maniatis T. (1989). Molecular cloning-A laboratory Manual, Cold Spring Harbor Laboratory Press, New York.

Lekha D., Mandaokar A.D. and Koundal K.R. (1999). Isolation and analysis of genomic clones of legumin storage proteins and their copy number determination in chickpea (*Cicer arietinum* L). Indian J. Expt. Biology 37: 1117–1122.

Shasany Ajit Kumar and Koundal K.R. (2000). Isolation and characeterization of legumin promoter sequence from chickpea (*Cicer arietinum* L). Indian J. Expt. Biology 38: 36–46.

12

Construction and Screening of cDNA Library

Complementary DNA (cDNA) is a double stranded DNA which is an exact copy of the mRNA, i.e. it consists of group of molecules possessing the information contained in various mRNAs but with termini equivalent to or related to those produced by restriction endonucleases. The use of cDNA is an absolute prerequisite for the expression of eukaryotic proteins in bacteria. The ability to synthesize and clone cDNA copies of mRNA has proved to be immensely powerful and versatile technique for study of gene structure, organization and expression. It has played a central role in the discovery of intervening sequences, RNA splicing and has provided the means of producing on a large scale polypeptides of academic, medical and commercial importance which had proved difficult by other means. A complete cDNA library is one that contains at least one cDNA alone representing each mRNA in the cell. There are about 36,790 individual types of mRNA molecules in a typical eukaryotic cell. The total number of clones, that should be screened (N) is given by

$$N = \frac{\text{In}\,(1\text{-}P)}{\text{In}\,(1\text{-}1/n)}$$

Where P = the probability of obtaining a given sequence

N = the minimum number of individual clones required

Using n = 36,790 as described above, then for a 99% probability of obtaining a given clone 1.69×10^5 independent recombinants should be screened.

i) cDNA Synthesis

Reagents

5 x MMLV reverse transcriptase buffer

Tris pH 8.3	=	250 mM
KCl	=	375 mM
DTT	=	50 mM
$MgCl_2$	=	15 mM

Protocol

a) 1st strand cDNA synthesis

1. Heat 10 µl poly A$^+$RNA (1µg/µl) at 65°C for 5 min. Place in ice.
2. When cooled to 0°C add
   a) 10 µl of 5 x MMLV RT Buffer
   b) 5 µl (1µg/µl) BSA (Final conc. 100 µg/µl)
   c) 2.5 µl dNTPs (10 mM each dNTP). Final conc. 500 µM of each
   d) 5 µl oligo $(dT)_{12\text{-}18}$. (500 µg/ml) final conc).
   e) 5 µl MMLV reverse transcriptase (1000 units) (Use 100 units of MMLV RT/µg of mRNA)
   f) 5 µl RNA sin (50 units)To the above add 10 µg denatured RNA (1µg/µl). Total Vol. = 50 µl
3. Mix gently
4. Transfer 10 µl aliquot to a separate tube containing 1 µl of a ^{32}P dCTP (1-10 µCi).
5. At time o' take 2 µl on DE 81 paper.
6. Incubate both tubes at 37°C for 1 hr.
7. At the end of reaction from the tube containing 10 µl take out 2 µl of aliquot on DE 81 paper and process as mentioned before.
8. Add 1 µl of 0.5 EDTA pH 8.0 to the rest 6 µl of reaction mix and store at -20°C to determine the yield of cDNA and to analyze it on alkaline agarose gel.
9. Place the tube containing 40 µl on ice and use for 2nd strand synthesis.

b) 2nd strand synthesis cDNA

10. In a separate tube add

7 µl	=	1 M Tris-Cl pH 8.3
28 µl	=	1 M KCl
1.4 µl	=	1 M $MgCl_2$
7 µl	=	10 mM d (NTP_s)
14 µl	=	100 mM DTT
16 µl (80 U)	=	Klenow fragment
1.7 µl (4U)	=	RNA_{se} H

2 µl (20 µCi)	=	^{32}P dCTP
202.9 µl	=	St.H_2O
280 µl	=	Total Vol.

11. Mix and add to 4 µl of DNA-RNA hybrid obtained at step 0. Mix gently.
12. Take out 4 µl each on DE 81 Paper. Keep one paper for processing and the other unprocessed to determine total counts.
13. Incubate the reaction mix at 16^{0}C for 2 hrs.
14. Take out 4 µl and spot on DE81 paper and process as done previously to determine incorporation.
15. Add 10 µl of 0.5 M EDTA.
16. Save 6 µl for running on alkaline gel.
17. Extract with phenol: chloroform (equal vol.)
18. Back extract phenol: chloroform layer with TE pH 8.0.
19. Extract aqueous phase with TE saturated ether. Use equal Vol.
20. To the aqueous phase add equal vol. of 4 M NH_4OAC and 2 Vol. of 100% chilled ethanol.

    Precipitation from ammonium acetate leaves short oligonucleotides in the supernatant, thus removing the oligo (dT) primer and enriching the pellet in longer cDNAs. Do not substitute sodium acetate for ammonium acetate.
21. Keep at -20^{0}C for overnight.

    Sample calculation for determining amount of cDNA synthesized.

    a) Assume 2 x 10^{12} cpm/Ci, thus, if [a-^{32}P] dCTP is at a specific activity of 3000 Ci/mmol, there are 6 x 10^{15} cpm/m mol.

    b) The second strand synthesis aliquot representing 1% of the total contains 10^5 TCA-precipitable cpm, thus

    $$\frac{10^5 \text{ cpm} \times 100}{6 \times 10^{15} \text{ cpm/mmol}} = 1.67\ 10^{-9} \text{ m mol } [\alpha\text{-}^{32}\text{P}] \text{ dCTP incorporated.}$$

    c) The initial concentration of [α-^{32}P] dCTP was

    $$\frac{50\,\mu\text{Ci/ml}}{3000\,\text{Ci/mmol}} = 1.67 \times 10^{-2}\ \mu\text{M}$$

    d) Thus, the total dCTP incorporated is

$$\frac{50\,\mu M}{1.67 \times 10^{-2}\,\mu\text{M}} \text{ x } 1.67 \text{ x } 10^{-9} \text{ m mol}$$

$$\times \frac{0.3\,\text{g dCMP}}{\text{m mol}} = 1.5\ \mu\text{g}$$

e) There were 4 dNTP in the reaction, so that the total cDNA synthesized in the second strandization is 6 µg x cDNA is double strandared. This is total 12 µg cDNA.

ii) Blunt-ending of double stranded cDNA

Protocol

1. Centrifuge the double stranded cDNA, kept for EtOH precipitation.
2. Wash the pellet with cold 70% EtOH. Dissolve the pellet in 20.5 µl of sterile water.

   (Take out 1 ml cDNA for native agarose gel electrophoresis. Add 9 ml of H_2O to it and keep at -20°C)
3. Complete second strand synthesis and for blunt ending the double stranded cDNA, add

   19.5 µl = cDNA in water
   0.5 µl = 10 mM dNTPs (all four)
   2.0 µl = 1 mg/ml BSA
   6.0 µl = 5 x T_4 polymerase buffer
   2.0 µl = (~ 10 units) T_4 DNA polymerase
   30.0 µl = St. H_2O

   Incubate at 37°C for 30 min.
4. Extract with equal vol. of phenol: chloroform (TE saturated). Back extract phenol: chloroform layer with equal Vol. of TE pH 8.0. Extract with ether the aqueous phase three times using equal volumes.
5. To aqueous phase, add equal vol. of 4 M NH_4 OAC; EtOH ppt. and keep at -20°C for overnight. Composition of 5 x T_4 polymerase buffer

	Stock (µl)	From	Final conc.
Tris acetate pH 7.9	100	1 M	330mM
KOAC	200	1 M	660mM
Mg OAC	30	1 M	100mM
BSA	6.75	2 mg/45 µl	1mg/ml
DTT	15	1 M	5mM
H_2O	248.25	To make up vol. to 600 µl	

iii) Agarose gel electrophoresis of double stranded cDNA

1 µl of second strand synthesized cDNA kept previously; run on 1.2% agarose gel in TAE.

Protocol

1. Weigh 0.72 g of agarose and add 60 ml of 1 x TAE

   Composition (50 X TAE)

Tris base	=	242 g
Glacial acetic acid	=	57.1 ml
0.5 M EDTA (pH 8.0)	=	100 ml
St. H_2O	=	to make one litre

2. Boil to dissolve agarose properly and then cool to 50°C. Cast the gel.
3. To 1 µl of cDNA add 12.5 µl of water and 1.5 µl of RNA loading buffer (10 X).
4. Load on gel. Load marker DNA also. Run at 50 V.
5. When bromophenol blue dye has reached the other end, remove the gel and stain in EtBr (1 µl/ml) for 10 min, destain with sterile water. Take the picture.
6. Dry the gel and expose to X-ray film for autoradiography.

iv) Methylation of cDNA

1. Spin down the cDNA for 20' in cold room in microfuge. Wash the pellet with 70% ethanol. Air dry the pellet. Dissolve in 22 µl of sterile water. And add.

2 x Methylase buffer	=	25 µl
100 mM - SAM	=	2 µl
EcoRI Methylase (20 U)	=	1 µl
Total	=	50 µl

2. Incubate at 37°C for 2 hr. At the end of reaction phenol: Chloroform extracts the mix. Back extract the P/C layer with $T_{10}E_1$ (pH 8.0), say 50 µl). Combine the aqueous phase and extract with saturated ether 3 X using equal volumes. Ethanol precipitate at -20°C for 2 hrs. and -70°C for 1 hr.

Note: The EcoRI methylation reaction protects the cDNA from EcoRI restriction when cohesive ends are generated on the synthetic linkers.

v) Ligation of *EcoR*I linkers to cDNA

Protocol

1. Centrifuge the EtOH ppt. from last step of previous experiment in cold room for 20' in microfuge at 4⁰C.
2. Wash the pellet with 70% ethanol. Air dry the pellet. Dissolve in 5.5 μl of st. H_2O.
3. To it add

10 x ligase buffer	=	1 μl
5 mM ATP	=	2 μl
EcoRI linker (0.5 mg/μl)	=	1 μl
T_4 DNA ligase (2weiss U)	=	0.5 μl
Total	=	10 μl

4. Incubate at RT for 1 hr. and then at 4⁰C for overnight

vi) *EcoRI* digestion after linker addition

Protocol

1. Centrifuge the ligation mix briefly.
2. Heat ligation mix at 65⁰C for 10' to inactivate ligase.
3. Keep on ice to cool and add the following

St. H_2O	=	100 μl
10 x *EcoRI* buffer	=	15 μl
BSA (1 mg/ml stock)	=	15 μl
EcoRI (18U/μl)	=	10 μl
Total vol.	=	150 μl

4. Incubate the mix at 37⁰C for 3-4 hrs.
5. Add 3μl (18 U/μl) of) of EcoRI and mix.
6. Incubate at 37⁰C for another 1 hr. or more.
7. Heat to inactivate the enzyme for 10 min at 65⁰C.

vii) Removal of excess linkers

During the incubation time prepare column of Sepharose CL-4B. The CL-48 column effectively removes EcoRI digested linkers that would otherwise interfare in subsequent cloning steps.

Reagents

i) Presswollen Sepharose CL-4B (Pharmacia) 4^{0}C

ii) CL-4B column buffer = 500 ml

1 M Tris-Cl pH 8.0	=	5 ml
5 M NaCl	=	60 ml
0.5 EDTA	=	1 ml
20% Sarkosyl	=	2.5 ml
St. Water to	=	500 ml

Protocol

1. Transfer 10 ml of sepharose slurry to a 50 ml polypropylene tube and fill with column buffer. Mix by inverting several times and let the sepharose settle gel, removing also the unsettled 'fines'.
2. Fill the tube a second and third time with column buffer. Allow to settle and remove the fines as above.
3. Add 10 ml of column buffer and mix by inverting several times. Incubate the tube at 37^{0}C for 10' and then proceed at RT. (Incubation done to remove bubbles).
4. Using the 1 ml st. pipette push a small piece of siliconized st. glass wool down to the tip of the 5 ml pipette Push a 3 cm length of plastic tubing firmly onto the tip of the 5 ml pipet. Clamp the tubing and attach the column to the stand.
5. With pipette carefully fill the column with the gel slurry from step 3. Release the clamp on the tubing and allow the column to flow. Periodically add more slurry until the vol. of packed gel in the column is at the 5 ml mark.
6. Wash the column 5-6 times with column buffer. Alow the level of buffer in the column to drop until it is just above the level of the gel, and clamp the tubing to stop flow.
7. After heat inactivation add 2 µl of 10 x gel loading buffer (without xylene cyanol) [$T_{10}E_1$+ Bromophenol blue is also fine] to the reaction mix and load the cDNA on the column.
8. Allow the loaded sample to enter the column, just until the top of the gel becomes dry. Collect 200 µl fractions manually. The cDNA may be followed with a hand held radiation monitor; the blue tracking dye indicates the position of digested linkers. Stop collecting before the blue dye begins to elute.

9. Count 2 μl aliquots of each fraction by spotting on DE 81 paper and dry it. Put in scintillation vial and add scintillation fluid based on tolune. Count and then plot on graph paper.
10. Pool the fractions with high counts and add 2.5 vol. of 100% EtOH (chilled). (Do not add NaCl as column buffer has sufficient NaCl).
11. Keep at -20°C for overnight and use for cloning after centrifugation.

viii) cDNA of cloning

Protocol

1. Set a water bath at 15°C.
2. Remove the gt^{10} arms and L buffer from the cDNA cloning system and allow thawing on ice.
3. Set up reactions on ice (Eppendorf) as described omitting T4 DNA ligase.

	Insert DNA	gt_{10} arms (μl)	Buffer (μl)	Water (μl)	T4 DNA ligase
1.	-	5 μl (0.5 μg)	1	4	2.5 U
2.	-	2 (1 μg)	1	6	2.5 U
3.	50ng linkered blunt end control DNA	2 (1 μg)	1	To 10, total vol.	2.5 U
4.	50ng cDNA	2 (1 μg)	1	To 10, total vol.	2.5 U
5.	100ng cDNA	2 (1 μg)	1	To 10, total vol.	2.5 U
6.	150ng cDNA	2 (1 μg)	1	To 10, total vol.	2.5 U

4. Mix gently and spin for a few seconds in a microfuge.
5. Add 2.5 U of T4 DNA ligase and mix gently.
6. Incubate at 15°C for 16-20 hrs (typically overnight).
7. Add 1 μl of 3M NaoAC and 27 μl 100% chilled ethanol. Keep at -20°C for 1 hr. and -70°C for 2 hrs.
8. Spin for 15 minutes in a microcentrifuge at 4°C.
9. Remove the supernatant and dry the pellet.
10. Carefully resuspend the pellet in 2.5 μl of $T_{10}E_1$

ix) Preparation of host bacteria

Protocol

1. Streak a loopful of the glycerol stock of each cell type (*E. coli* L87 and *E. coli* NM514) supplied onto separate dry L-agar plates. Incubate overnight at 37°C.

2. Streak a loopful of the glycerol stock of each cell type (*E. coli* L87 and *E. coli* NM514) supplied onto separate dry L-agar plates. Incubate overnight at 37ºC.
3. Add 1 ml of each overnight culture to 50 ml of prewarmed L-broth + 0.4% maltose. Incubae the cells at 37ºC with vigorous shaking until they reach on OD_{600} of 0.5 (2.5 x 108 cells/ml). This will take approximately 3 hours.
4. Cool the culture on ice and spin at 3000 rpm for 10 minutes at 4ºC.
5. Resuspend the 15 ml of ice-cold 10mM $MgSO_4$ and mix thoroughly.
6. Store at 4ºC. The cells are now ready for infection with phage.

Reagents

Protocol

x) *In vitro* packaging

Protocol

1. For each packaging reaction to be performed remove one blue tube (extract A) and one yellow, tube (extract B) from the packaging reaction box stored at -70ºC; place both tubes in ice immediately.
2. Allow contents of each tube to thaw on ice.
3. While extracts are thawing, make sure that two pipettes set at appropriate volume are available for instant use, one for extract A.
4. Add 15 µl from blue tube to the ligation reaction.
5. Immediately transfer 15 µl of the contents of a yellow tube into the reaction tube and mix gently using the tip of the pipette used for the transfer.
6. Centrifuge the reaction tube containing the complete reaction mix for 10 seconds.
7. Incubate in a 20ºC water bath for 2 hours.
8. Add 0.5 ml SM buffer followed immediately by 10 µl $CHCl_3$. Mix gently.

xi) Titration of λ gt10 recombinants

Protocol

1. Add 30 µl of the final packaged phage mixture to 27 µl of SM buffer (10^2 dilutions). Mix well.
2. With a fresh tip add 30 µl of the 10^2 dilution to 270 µl of SM buffer (10^3 dilutions). Mix well.
3. Repeat step 2 until the 10^7 dilution.

4. Prepare a duplicate dilution series starting with a fresh 30 µl sample of the final packaged phage mixture.
5. Melt top agar and place in a 45^0 C water bath.
6. Prepare a series of tubes for each of the two cell types (L87 and NM 514) by aliquoting 100 µl of each type of plating cells into a suitable container lavelled clearly with the cell type and proposed dilution.
7. Add 100 µl of each dilution (10^3 to 10^7) into the appropriate tube for both the L87 and MN514 series of cells.
8. Incubate at 37^{0}C in a water bath for 15 minutes.
9. To each dilution add 4 ml of liquid top agar at 45^0 C. Mix quickly pour onto a dry 90mm plate containing approximately 20 ml of L-agar.
10. Allow top agar to set completely before moving the plates (10 minutes).
11. Incubate the plates (upside down) at 37^0 C overnight.
12. Count the number of plaques on those plates with upto 1000 plaques.
13. The phage litre per ml is simply calculated by multiplying the numbers of plaques by the dilution number. For example, 150 plaques on the 10^4 dilution plate represents a titre of 1.5 x 10^6 plaque forming units (pfu)/ml.

xii) Amplification of the cDNA library

Protocol

1. Plate out phage on NM 514 cells at 3 x 10^4 recombinants per 90 mm plate.
2. Incubate at 37^{0}C until the plaques appear and merge.
3. Place the plates at 4^{0}C and add 5 ml of SM buffer per plate. Leave at 4^{0}C overnight.
4. Remove SM buffer, wash plate with 1ml of fresh SM buffer. Add $CHCl_3$ to a final concentration of 5%. Incubate at RT for 15 minutes.
5. Remove cell debris by centrifuging for 5-10 minutes at 2000g.
6. Recover supernatant and transfer to a sterile polypropylene or glass bottle. Add $CHCl_3$ to 0.3% and store in aliquots at 4^{0}C.
7. Add DMSO to 7% and store at -70^{0}C for longer storage.

xiii) Screening of λ gt10 cDNA Library

Protocol

In order to isolate a recombinant cDNA clone which encodes for a specific mRNA sequence, the usual approach is to screen the cDNA. If the selected clone confers some phenotypic character in the bacterial host it is easy to select them by giving appropriate selection conditions. But majority of eukaryotic genes or even bacterial genes do not encode a gene with a selectable function. Hence, they have to be identified by screening libraries based on any one of the approaches i.e. (i) hybridization to a specific nucleic acid probe, (ii) by using antibodies against a particular protein or (iii) by ampliofying a specific sequence using defined set of primers.

To screen a library using a nucleic acid or antibody as a probe a large number of clones have to be screened at one time. For doing this the library is spread out on agarose, then the clones are transferred to nitrocellulose or nylon membrane, filters are then processed and then hybridized with appropriately labeled nucleic acid or antibody probe. The desired clones are then picked up and purified.

The number of clones that should be screened to get a desired clone depends upon the abundance of the mRNA in the cell. The amount of protein in the cell for the sequence one is looking for is a good indication of abundance of the mRNA. For example if the proteins that comprise 1% of the total cell protein are made by mRNA that usually comprise 1% of the total poly (A^+) RNA, the desired cDNA clone should comprise about 1% of clones in cDNA library.

xiv) Planting bacteriophage library and transfer of phage to membrane filters

Reagents

1. Host bacteria.
2. Recombinant library.
3. Membrane filters (62 mm).
4. 3 Sheets of Whatman 3 MM paper (46 x 57 cm each).
5 0.2 M NaOH, 1.5M NaCl (denaturing solution).
6. 0.4 M Tris (pH 7.6, 2x SSC.
7. 2xSSC diluted from 20 x SSC stock (20xSSC: 3 M NaCl, 0.3M Sodium citrate, pH 7.0).
8. 0.7% top agarose.
9. LB agar plates.

Protocol

1. Determine the titre of the library by serial dilution as described earlier.
2. Mix recominent phage and plating bacteria in culture tube and incubate for 20 min at 37°C (These steps are described in earlier section on tittering phage library).
3. Add 0.7% top agarose and pour on calculated number of LB agar plates. Incubate for 6 to 12 h at 37°C until plaques cover the plate but are not confluent. Place at 4°C for 1 h before applyijng filters.

   Note: The top agarose must be melted and cooled to 45-50°C before use. If it is too hot and bacterial cells will be killed and if it is cold and agarose will solidify in the tube. Because small plaques are preferred, do not incubate unattended overnight at 37°C. When the plaques start forming keep them at 4°C and continue growth next day till the desired size of the plaques are formed.
4. Label the memberane filters with water proof ink. Gently place the filter (with ink side up) on one edge of the plate and slowly lay down progressively without incorporating air bubbles under it or disturbing the gel surface.

   Note: The filters once kept should not be repositioned.
5. Leave the filters on the plate for 1-10 min at room temperature. Mark orientation of the filter on the plate for 1-10 min at room temperature. Mark orientation of the filter to the plate by piercing with 20 G needle through the filter into the agar asymmetrically at different points around the edge of the plate. These points are also marked on the bottom of the plate with a permanent marker. Using flat and ended forceps, slowly lift the filter from the plate and place plaque side up on a filter paper for air drying for 10-20 min at room temperature.
6. Take 3 trays and put Whatman 3 MM sheets in them.

   Add to the trays in sequence – Tray 1: 0.2 M NaOH 1.5 M NaCl, Tray 2: 0.4 M Tris-HCl (pH 7.6), 2x SSC, and Tray 3: 2 x SSC; to let the filter paper in them get saturated.
7. Incubate the filters with plaque side up sequentially for 1-2 min in each of the trays.
8. Finally dry the filters in a vacuum over for 90-120 min at 80°C or overnight at 42°C in an oven without vacuum.

9. The dried filter can be wrapped in dried filter paper on plastic wrap and stored at room temperature until used for hybridization.
10. Do the hybridization and screening as described for genomic library.

Suggested reading

Benton W.D. and Davis R.W. (1977). Screening lambda gt recombinant clones by hybridization to single plaques *in situ*. Science 196: 180-183.

Datta Subhojit, Rekha Kansal and Koundal K.R. (2000). Construction and screening of cowpea (*Vigna unguiculata* L.) cDNA library for lectin genes. J. Plant Biochem. Biotech. 6: 67–71.

Grunstein M. and Hogness D.S. (1975). Colony hybridization: A method for the isolation of cloned DNAs that contains a specific gene. Proc. Natl. Acad. Sci. USA 72: 3961-65.

Okayama H. and Berg B. (1982). High efficiency cloning of full length cDNA. Mol. Cell Biol. 2: 161-170.

Old R.W. and Primrose S.B. (1994). Principles of Gene Manipulation. Blackwell Scientific Publication, London.

Sanghamitra Saha, Rekha Kansal and Koundal K.R. (1996). Use of polymerase chain reaction in analysing recombinant cDNA clones encoding storage proteins of chickpea (*Cicer arietinum* L). Ind. J. Expt. Biol. 34: 1019–1025.

13

PCR and RT-PCR Methods

The polymerase chain reaction (PCR) is a very powerful new method used for *in vitro* to amplify a deoxynibonuclic specific region that lies between two regions of known DNA sequence single molecule to million folds in short time. This is achieved by an enzyme *taq. DNA polymerase* which is thermostable using this enzyme can synthesize microgram quantities of DNA of specified length and sequence from pictogram quantities of complex template in a few hours time. This technique first described by saiki *et al* (1985) has lot of applications in various fields of molecular biology, especially in studies of genes which are expressed poorly and in low amounts and in RFLP mapping or study of DNA polymorphism. The *in vitro* DNA amplification by polymerase chain reaction is based on two oligonucleotide primers which flank the specific DNA fragment to be amplified and hybridized to opposite strands. The PCR technique of amplification involves three steps: Denaturation of the double stranded DNA fragment to be amplified and hybridized to opposite strands. The PCR technique of amplification involves three steps: i.Denaturation of the double stranded DNA template at high temperature (94°C), ii. Annealing of the oligonucleotide primers to the DNA template at low temperature (37° - 55°C), iii.Extension of annealed primer useing taq DNA polymerase (Klenow fragment at 72°C).

These three steps from one cycle, repeated cycles of heat denaturation of the template, annealing of the primers and extension of the annealed primer with polymerase results in the amplification of the template DNA defined by the 5' end of the two primers. The extension product of one primer serves as the template for the other primer and each cycle doubles the amount of DNA produced in the previous cycle and results in the exponential accumulation of the DNA template in several million folds in a short time. The main advantage of this technique is in amplification of gene sequence. Low abundance mRNA can be amplified using this technique. For this, after isolating the low abundance mRNA its complementary DNA is synthesized using reverse transcriptase. The cDNA synthesized by this process can then be amplified by PCR and used for further studies. Tm (°C) = 4(G+C) + 2(A+T). Primer annealing temperature depends on its Tm value.

Note: PCR is capable of amplifying as little as single molecule of DNA, therefore, precaution should be taken to guard against contamination of the reaction with the trace amount DNA that could serve as template.

A. PCR- Polymerase chain reaction

i). DNA amplification by PCR

Reagents

1. 10 x reaction buffer 100 mM Tris-HCl (pH 8.3), 500 mM KCl, 15mM $MgCl_2$
2. dNTP mix 200 µM each dNTP (dCTP, dGTP, dTTP in water).
3. Primers = 0.6 µg/µl
4. Template DNA = 0.5 µg/µl in TE
5. Taq polymerase 5 u/µl (dilute to 0.5 u/µl before use)

Protocol

1. To a 0.5 ml microfuge tube add the following in order :

   Sterile Milli-water = 14.5 µl

   10 x Reaction Buffer = 2.5 µl

   dNTP mix = 4.0 µl

   Primer 1 = 1.0 µl

   Primer 2 = 1.0 µl

   Template DNA = 1.0 µl

   Taq polymerase 5u/µl. Vortex briefly, then spin down for 3 seceond
2. Overlay the reaction mixture with 60 µl silicon oil to prevent evaporation during repeated cycles of heating and cooling.
3. Carry out amplication with the following programme: 35 cycles each of denaturation at 94°C for 1.5 min, annealing at 55°C for 2 min and extension at 72°C for 2 min. After the last cycle incubate the sample for an additional 7 min at 72°C.
4. Remove 5 µl at the amplified DNA from the tube and analyse on a 10% polyacrylamide gel or on a 2% mini agarose gel after adding 1 µl of 5 x loading dye.
5. View gel under uv light to analyse.

ii) Purification of PCR fragments for cloning

After an aliquot of the PCR mixture is analyzed on an agarose gel, the remainder of the reaction is concentrated by ethanol precipitation, resuspended in buffer, and subjected to a simultaneous fill-in/kinase reaction with the Klenow fragment of *E.coli* DNA polymerase and T-4 polynucleotide kinase, the four deoxynucleotides and rATP. The reaction then is loaded onto a preparative 1, 1.5, or 2% low-melting temperature agarose gel, depending on the size(s) of the fragment(s) as determined above, and after minimal separation is achieved between the product(s) and the excess primers. The DNA fragments are exercised and eluted. After concentration by ethanol precipitation, amplified DNA fragments are ligated into blunt-ended cloning vector, such as Smal-linearized, dephospohorylated double-stranded M13 replicative form of pUC.

It should be noted that several other methods to purify the DNA fragments for cloning have been investigated. These included standard ethanol-acetate precipitation, a 50% ethanol precipitation, spin-column purification and precipitation with polyethelene glycol (PEG). The first three methods did not remove sufficient unincorporated primer, and, during the subsequent ligation of the DNA fragment, the primers apparently competed for the blunt ended vector during ligation because the efficiency of ligation was significantly lower and the vast majority of recombinant clones contained only primer derived inserts. Precipitation by polyethylene glycol resulted in only an extremely small DNA pellet, removing the PEG supernatant is difficult, and yield of PCR product were variable.

Protocol

1. 1 Precipitate the PCR reaction by addition 2.5 volumes of 95% ethanol containing 0.12 M sodium acetate, pH 4.8.
2. Perform the combined fillin-kinase reactions previously described.
3. 3. Add 5 μl of agarose gel loading dye and load the reaction into a well of a 1.0% low-melting temperature agarose gel. Electrophorese for 30-60 minutes at 100-120 mA, and then excise the desired band visualized under UV light with a clean razor blade.
4. Elute the DNA from the gel by standard freeze-thaw methods, followed by a phenol extraction and concentrate by ethanol precipitation.
5. Resuspend the dried DNA in 10 μl of 10:0.1 TE buffer. Use this DNA in a standard blunt-ended ligation reaction. Typically, use 2-3 μl of this DNA in a 10 μl ligation reaction with 20 μg of pUC 18/Smal-CIAP, although this will depend on the yield of amplified DNA from the PCR reaction.

B. RT-PCR –Reverse transcription-polmerase chain reaction

Reverse transcription polymerase chain reaction (RT-PCR) is a laboratory technique combining reverse transcription of RNA into DNA (in this context called complementary DNA or cDNA) and amplification of specific DNA targets using polymerase chain reaction (PCR). The quality and purity of the RNA template is essential for the success of RT-PCR. The first step of RT-PCR is the synthesis of a DNA/RNA hybrid. Reverse transcriptase also has an RNase H function, which degrades the RNA portion of the hybrid. The single stranded DNA molecule is then completed by the DNA-dependent DNA polymerase activity of the reverse transcriptase into cDNA. The efficiency of the first-strand reaction can affect the amplification process. From here on, the standard PCR procedure is used to amplify the cDNA. The possibility to revert RNA into cDNA by RT-PCR has many advantages. RNA is single-stranded and very unstable, which makes it difficult to work with. It is primarily used to measure the amount of a specific RNA. This is achieved by monitoring the amplification reaction using fluorescence, a technique called real-time PCR or quantitative PCR (qPCR).

Reverse Aid H minus First strand cDNA synthesis kit from Fermentas is used for first strand cDNA synthesis

Components provided in the kit:

1. Reverse Aid H minus M-Mulv Reverse Transcriptase (200units/ml)
2. Ribonuclease inhibitor (200units/ul)
3. 5X Reaction buffer (250 mM Tris-Cl, pH 8.3, 250 mM KCl, 20 mM $MgCl_2$, 50 μM DTT)
4. 10 mM dNTP mix
5. Oligo dT primer mix
6. DEPC treated water

i) Synthesis of first stand cDNA by reverse transcriptase suitable for PCR

Protocol

1. Prepare the following reaction mixture in a Eppendorf tube on ice :

   Template RNA, the amount of total RNA or poly $(A)^+$ RNA required for reaction depends on the level of expression of a gene

Total RNA	10 ng - 5 μg
or poly$(A)^+$RNA	1 ng - 0.5 μg

or Specific RNA	0.01pg – 0.5ug
Oligo $(dT)_{16}$ primer (0.5 ug/ul)	1 ul
DEPC treated water to	12 ul

Mix gently and spin down for 30 sec in micro centrifuge.

2. Incubate the mixture at 70°C for 5 min, chill on ice and collect drops by brief centrifugation.
3. Place the tube on ice and add the following components in the indicated order :

5X reaction buffer	4 ul
Ribonuclease inhibitor (20 units/ul)	1 ul
10 mM dNTP mix	2 ul

Mix gently and collect drops by centrifugation

4. Incubate at 37°C for 5 min. add
5. Reverse Transcriptase (200units/ul) 1 ul
6. Incubate the mixture at 42°C for 60 min
7. Stop the reaction by heating at 70°C for 10 min. Chill on ice the first strand cDNA synthesized can be used directly for amplification by PCR.

ii) PCR Amplification of second strand cDNA

Reagents

1. 10XPCR buffer
   30 mM dNTP mix (7.5mM each dNTP)
2. Gene specific Forward and Reverse Primers (100 ng/ul)
3. Taq DNA polymerase

Protocol

1. Add the following in a 0.2 ml PCR tube

First strand cDNA mix	2 ul
10XPCR buffer	5 ul
30 mM dNTP mix	1 ul
Forward primer	1 ul
Reverse primer	1 ul
1UnitTaq polymerase	1 ul

Nuclease free demonized water to final volume of 50ul. Keep a control reaction tube without DNA template.

2. Mix the contents and spin down quickly and perform the reaction in a Thermal cycler with the following standardized programme

   Initial denaturation at 94°C for 4 min, then do 30 cycles of the progrmme scheduled as follows:

Denaturation at 94°C	45 sec
*Annealing at 52°C	30 sec
Extension at 72°C	1 min
Final extension at 72°C	10 min

   *Annealing temperature will vary with Tm of the primers.

3. After the amplification reaction the amplified product is checked on 1% agarose gel along wit 1kb DNA ladder.

Note: Ribonuclease is difficult to inactivate. Hence, it is very important that utmost care is taken to avoid contamination. Gloves should be born all the times, use sterile disposable plastic ware and bake glass ware at 200°C overnight; are some of the steps which can help to prevent contamination by RNase while performing the experiments of RNA isolation and processing.

Suggested reading

Mullis K., Faloona F., Scharf S., Saiki R., Horn G. and Erlich H. (1986). Specific enzymatic amplification of DNA in vitro: the polymerase chain reaction. Cold Spring Harbor Symp, Quant Biol 51: 263-273.

Oyce C. (2002). Quantitative RT-PCR. A review of current methodologies. Methods Mol. Biol. 193: 83–92.

Saiki R.K., Gelfand D.H., Stoffel S., Scharf S.J., Higuchi R., Horn G.T., Mullis K.B. and Erlich H.A. (1988). Primer directed enzymatic amplification of DNA with thermostable DNA polymerase. Science 239: 487-491.

Sanghamitra Saha, Rekha Kansal and Koundal K.R. (1996). Use of polymerase chain reaction in analysing recombinant cDNA clones encoding storage proteins of chickpea (*Cicer arietinum* L). Indian. J. Expt. Biol. 34: 1019-1025.

14

M13 Cloning for Di-deoxy DNA Sequencing

M13 is a single stranded filamentous DNA phage. Vectors have been developed from M13 phase by Joachim Messing (1983), which are suitable for preparation of single stranded DNA required for sequencing by Sanger's method. These are derivatives of M13 wild type (size 6.407 kb). The single stranded phage on entering a suitable host (*E.coli* F) is converted to double stranded RF forms. This process proceeds until about 100-200 copies are made. During RF replication, gene 5 products accumulate. When the amount of this rises above a critical limit, gene 5 product by a mechanism not understood, acts together with gene-2 product (nicks the DNA near the origin) to cause some of the RF (replicated form) to switch to rolling circle of replication of the + strands (infectious viral DNA strand). This results in the production of single strandd DNA circles which are packaged into phage particles and secreted from cells without lysis. The supernatant of such culture of infected cells has single strand of phage DNA.

i) Preparation of phage DNA (RF DNA)

Solutions

1. 3M Potassium acetate — 10.0 ml

   5 M Potassium acetate — 6.0 ml

   Glacial acetic acid — 1.15 ml

   DDW — 2.85 ml

   Make up as directed in order given

2. 0.2 M NaOH, 1% SDS solution

   Sodium hydroxide — 0.08 g

   SDS — 0.10 g

   DDW — 10.00 ml

3. LB medium (1000 ml)

   Bacto-tryptone — 10.0 g

Yeast extract	5.0 g
Sodium chloride	5.0 g
D-glucose	1.0 g
DDW	1000 ml

Make up as directed in a flask and autoclave

4. 2x YT medium

Bactotryptone	16.0 g
Yeast extract	10.0 g
Sodium chloride	5.0 g
DDW	1000 ml

Make up as directed and autoclave

Protocol

1. Put 10 ml of LB medium in a small bottle. Incubate with one colony of JN 103, JM107/JM109 cells from an N9 minimal agar plate. Grow with shaking at 32°C to an OD at 600 nm of 1.0. Store these cells at 4°C.
2. Place 10 ml of LB medium in the second sterile bottle. Inoculate this medium with one blue plaque from the mp vector plate by transferring a blue plaque in an agar plug in the tip of a sterile Pasteur pipette. Grow with shaking at 37°C overnight to make a plaque stock.
3. Add 100 μl of LB containing JM103 from Step 1 to 10 ml of 2 x YT medium. Grow with shaking at 37°C to an OD at 600 nm of 1.5.
4. Add 1 ml of plaque stock from Step 2 to these cells and infect by shaking at 37°C for 15 min.
5. Add 0.1 ml of chloramphenicol solution to the infected JM 10^8 cells. Shake at 37°C for 10-12 hrs to allow accumulation of RF DNA.
6. Spin down the culture and dry the pellet.
7. Add 400 μl of SDS/NaOH solution. Keep in ice for 5 min.
8. Add 300 μl of 3M potassium acetate. Keep in ice for 3 min and then centrifuge. Transfer 0.7 ml of the supernatent to the other Eppendorf tube.
9. Add 0.7 ml of phenol-chloroform (50:50), mix well and centrifuge. Take again 0.7 ml of the aqueous layer and add equal vol of cold chloroform: isoamylalcohol (24:1). Mix well and spin. Take 0.5 ml of the aqueous layer.
10. Add 1 ml of 100% chilled ethanol and keep at-20°C for 1 hr for DNA precipitation. Spin to pellet the DNA. Wash the pellet with 70% chilled ethanol and dissolve in 50 μl TE.

ii) M13 Cloning

Solutions

1. TE buffer (10 mM Tris, 1mM EDTA, pH 7.5 (1000 ml)

Tris	1.21 g
EDTA	0.372g
Sterile Distt. H_2O	1000 ml

Autoclave before use

2. 5 x high salt buffer (500 mM NaCl, 250 mM Tris, 50 mM $MgSO_4$ pH 7.4)

5 M NaCl	0.10 ml
1 M Tris	0.25 ml
1 M $MgSO_4$	0.05 ml
St DDW	0.60 ml

Make up as directed in order. Autoclave before use.

3. 10 x ligase reaction buffer (1 ml)

2M Tris pH 7.8	0.25 ml
1M $MgCl_2$	0.10 ml
1 M DTT	0.20 ml
ATP	5.5 mg
BSA Soln, (5mg/ml)	0.1 ml
Sterile H_2O	0.35 ml

Protocol

Preparation of vector M13 mp 18 RF DNA and DNA fragments

1. Set up the restriction as follows in an Eppendorf tube:

Samples	DNA	EcoRI (10 units)	High salt buffer	H_2O	Super-mine	RN Ase	Total (Vol.)
M13 mp 18 DNA (5µg)	5 µl	2 µl	4 µl	10 µl	-	-	21 µl
Insert DNA (5µg)	10 µl	2 µl	4 µl	4 µl	1 µl	1 µl	22 µl

2. Incubate at 37°C for 3 hr. Terminate by incubating at 65°C for 5 min. Extract both the restricted DNAs by phenol/chloroform and ethanol precipitate and dissolve the final pellet in 20 µlTE.

Ligation of DNA fragments into vector DNA

1. Set up the ligation reaction as follows:

M13 mp18 (cut vector) DNA	2 μl
10xligase reaction buffer	1 μl
DNA insert (cut)	1 μl
T4 DNA ligase enzyme	1 μl
Sterile distilled water	5 μl

2. Incubate at 15°C for 12 hr or at room temperature for 2 hr. Dilute with 0.1M $CaCl_2$ to 100 μl. Use 10 μl for transformation and store rest at -80⁰C.

Preparation of Competent Cells

	L-Broth	M9 minimal medium	
Bactotryptone	10.0 g	KH_2PO_4	5.0 g
Yeast extract	5.0 g	NH_4Cl	1.0 g
Thymidine	20 mg	NaCl	0.5 g
1N NaOH	1.0 ml	DDW to	1000 ml
NaCI	5.0 g	Na_2HPO_4	6.0 g
DDW	1000 ml		
Autoclave			

Protocol

1. Streak out the host *E.coli* JM 107 on M9 minimal agar plates and grow overnight at 37°C.
2. Pick one colony and grow overnight in 5 ml of L-Broth at 37⁰C, with constant agitation.
3. Dilute inoculum 1:100 into L-Broth and grow at 37⁰C, until an OD of about 0.5 (at 660 nm) is reached.
4. Chill the cells on ice. Prechill all solutions, pipettes, etc. for the remainder of the protcol.
5. Spin JM 107 culture at 4K for 3 min. in Sorvall Centrifuge.
6. Resuspend pellet in 1/5th of the growth vol of CM_1 buffer (10mM NaOAC pH5.6, 50mM $MnCl_2$, 5mM NaCI). Place on ice for 20 min.
7. Centrifuge at 3K for 2 min.
8. Resuspend the pellet in 1/50th of the growth volume in CM_2 buffer (10mM NaOAC pH 5.6, 5% glycerol, 70mM $CaCl_2$, 5mM $MnCl_2$).
9. Place in 100 μl aliquots and store at -70⁰C.

Note: Stored competent cells can be used upto a period of six months.

iii) Preparation of plating cells and transformation

i) Set up overnight host strain in 10 ml of L-Broth containing 0.2% maltose.

ii) Spin down the overnight cultures on the bench top centrifuge for 15 min.

iii) Pour off the supernatant. Resuspend the pellet in 3.3 ml of age buffer pH 7.5.

iv) Store the preparation in fridge until needed. It is viable for about one week.

Transformation

a) 0.1M IPTG (i.e., 2.38 mg/0.1 ml of dist. H_2O).

b) 2% X-gal (i.e., 8 mg/0.4 ml of dimethyl formamide) can be stored at -20^{0}C in dark.

1. Take 10 µl of diluted ligated M13 DNA in an Eppendorf tube.
2. Add 100 µl of competent cells.
3. Keep on ice for 2 min. Heat shock at 42^{0}C for 2 min and incubate at 37^{0}C for 4 min.
4. Add 200 µl of plating cells. Mix thoroughly and add the followings:

10 µl	0.1 IPTG
50 µl	2% X -gal
3 ml	top layer agar at 55^{0}C.

Mix well and pour onto YT agar plates. The transformants with inset will form colourless plaques. The blue plaques represent M13 mp18 DNA.

iv) Preparation of single stranded DNA from M13 recombinants

1. Dilute overnight grown culture of JM 107 into 100 ml of 2 X YT broths. Use this diluted culture 2 ml/plaque.
2. Using a pasteur pipette, blow the plug of agar containing white plaque (recombinant) into the 2 ml of 2 X YT inoculated media.
3. Incubate overnight with shaking at 37^{0}C.
4. Pour off 1.5 ml of overnight grown culture into 1.5 ml Eppendorf and spin for 5 min. (Pellet will contain the double stranded DNA. Process it if you like as described earlier).
5. To 1 ml of supernatant obtained add 250 µl of 20% PEG 6000 (or 100 µl of 50% PEG 6000) and 225 µl of 2.5M NaCl. Mix well and leave for 20 min. at room temperature.

6. Centrifuge for 5 min. Discard the supernatant and take the pellet and centrifuge again for 10 sec. Remove excess PEG 6000 with a syringe.
7. Resuspend the pellet in 100 μl TE buffer and add 50 μl phenols. Vortex for 15 sec. and centrifuge for 3 min. Remove the aquous phase.
8. Add 50 μl of chloroform to the acquous phase and again spin for 3 min.
9 Take out the upper layer in a separate Eppendorf tube, add 10 μl of 3M sodium acetate, 300 μl of 100% ethanol. Keep at -20°C for 1 hr.
10. Spin for 10 min., wash the pellet with 70% ethanol and dry the pellet under vacuum. Dissolve the pellet in 20 μl TE and store at -20°C.

v) Polyacrylamide gel electrophoresis of recombinant DNA

Reagents

1. 10 x TBE buffer, pH 8.3 (1000 ml)

Tris	=	109.0 g
Boric acid	=	55.0 g
Na_2EDTA	=	9.3 g

Make up to 1 litre with st. distilled water

2. 50 x TAE = 108.0 g

Tris	=	24.2 g
Na_2EDTA	=	3,72 g
Glacial acetic acid	=	5.71 ml
Distt. water	=	100 ml

3. Ethidum bromide solution = 10 ml

Ethidium bromide	=	100 mg
Distilled H_2O	=	10 ml

No need to autoclave, store until needed

4. 10 x Gel running dye = 0.1 ml

Glycerol	=	2.5 ml
50 x TAE	=	1 ml
0.5 M EDTA pH 8.0	=	1 ml
Bromophenol blue 0.5%	=	1 ml
Ficoll	=	500 mg

5. Formamide dye mix: 0.1% (w/v) xylene cyanol FF, 0.1% (w/v) bromophenol blue, 10mM Na_2EDTA, 95% (v/v) deionized formamide. Store at 4°C, make fresh monthly.

Protocol

1. Prepare a horizontal 1% agarose mini gel. To 0.6g of agarose, add 50 ml of water. Heat until dissolved. Then add 6 ml of 10 x TBE buffer. Make upto 60 ml final volume with sterile water. Add 6 μl of ethidium bromide stock solution. Cool the gel to 37^{0}C befor pouring. This will prevent splitting.
2. Place the gel in the tank. Pour appropriate amount of TBE electrophoresis buffer, containing 1 μg/ml ethidium bromide.
3. Mix 2 μl of prepared recombinant DNA with 4 μl of running dye. Pipette the samples into the slots taking great care. Change the tip between each operation. Carry out electrophoresis and view gel under UV light.

Suggested reading

Messing J. (1983). New M13 vectors for cloning. Methods in Enzymology 101: 20-78.

Sambrook J., Fritsch F.F. and Maniatis T. (1989). Molecular cloning-A laboratory Manual, Cold Spring Harbor Laboratory Press, New York.

Yanisch-Perron C., Viera J. and messing J. (1985). Improved M13 phage cloning vectors and host strains: Nucleotide sequences of the m13mp18 and puc19 vectors. Gene 33: 103-199.

15

DNA Sequencing Methods

DNA can be sequenced in both double stranded and single stranded templates by cloning in plasmid or phage vectors. There are two sequencing methods.

A. Sanger's method

B. Maxam Gilbert Method

A. Sanger's Method

The dideoxy chain terminator DNA sequencing procedure is considered one of the most simple, rapid and accurate methods of DNA sequencing. It makes use of the ability of DNA polymerase to faithfully synthesize a complementary radioactive copy of a single stranded (SS) DNA template using a short DNA fragment as primer. Once the dideoxynucleotide is incorporated, the 3' end is no longer a substrate for chain elongation and the growing DNA chain is terminate.

The large fragment of DNA polymerase-1(Klenow enzyme) which lacks 5'-3' exonuclease activity of the intact enzyme is used to synthesize a complimentary copy of the SS target sequence from an adjacent annealed primer. The reaction is carried out in the presence of the four deoxynucleoside triphosphates one or more of which is ^{32}P labelled and inturn with each of the four dideoxy nucleoside triphosphate in four separate incubations. Therefore, in each reaction there is a population of partially synthesized DNA molecule each radiactive, having a common 5'-end but each varying in length, to a base-specific 3' end. Parallel size fractionation of the four DNA replication reaction products by denaturing gel electrophoresis and autoradiography allows a sequence to be deduced.

The DNA fragment (DS DNA) to be sequenced is inserted into polylinker located in the non-essential region of the phage genome. M13 mp vectors have lac Z gene which makes the screening of recombinant easy using IPTG and X-gal. Thus DNA to be sequenced is cloned in M13 mp 18 RF DNA, transformation done and single stranded recombinant DNA is obtained as described earlier in M13 cloning.

i) Annealing of M13 primer to template

To prepare material for the four sequencing reaction s (one clone)

Single stranded DNA template	5 μl
M13 Primer (4 pg) 15bp	2 μl
10 X polymerase reaction buffer-	1.0 μl
DDW	4.5 μl

Mix the components well and spin briefly to the bottom of the tube. Incubate at 85-90^{0}C for 2 min and allow to cool slowly still sitting in the water bath.

ii) Sequencing reaction

1. Take the annealed primer/template, Spin briefly to the bottom 12.5 μl
2. Add polymerase (2 units/ μl) 1 μl

   DTT (0.1 M) 1 μl

   ^{35}S-ATP/^{32}P-alpha-ATP 15.5 μl

   Vortex to mix; spin briefly to bottom of tube. This is now the template/primer/label/ enzyme mix
3. Now take four microcentrifuge tubes, mark them A,C,G and T (four tubes per DNA sample to be sequenced). To each tube carefully add 3 μl of annealed/template/primer/label enzyme mix. (Enzyme mix). Prepare sequencing reaction mix as given earlier
4. To each tube carefully add 2 μl of the relevant dNTP/ddNTP mixtures. Add each droplet separately just inside the rim of the tube.

i.e. enzyme mix + 2 μl A^{0}/ddATP	=	Tube A
enzyme mix + 2 μl C^{0}/ddCTP	=	Tube C
enzyme mix + 2 μl G^{0}/ddGTP	=	Tube G
enzyme mix + 2 μl T^{0}/ddTTP	=	Tube T
5. Spin briefly to mix and incubate at 30°C for 20 min.
6. Add 1 μl of chase NTP the 0.5 mM dATP to each of the four reaction tubes. Mix with a brief spin and incubate at 30^{0}C for 15 min.
7. If samples are to be electrophoresed immediately, add 4 μl formamide dye, mix to the rim of each tube and spin briefly.
8. Place tubes in a 90^{0}-95^{0}C water bath for at least 3 min. Load 2 μl from each tube immediately onto sequencing gel. If samples are not to be electrophoresed immediately store at -20^{0}C for not longer than 24 hrs.

iii) Dideoxy reactions

1. While the DNA/primer is annealing, label 400 µl Eppendorfs by sample and nucleotide A, C, G and T. Add to each tube the appropriate N°/ddN mix 2 µl near the bottom of the tube, i.e., A gets 2 µl of A mix., tube C gets 2 µl of C mix, etc. from step 4.
2. Cut off tops of the tubes and place in microfuge racks in order (A.C.G.T.).
3. After completion of the annealing incubation prepare reaction mixtureIn a 1.5 ml tube add:1 µl per sample plus 1 V of 0.1 M DTT (dithiothreitol1 µl per sample plus 1 µl of 1 unt/ µl Klenow large fragment in dilution buffer1 µl per sample plus 1 µl of ^{35}S dATP(Klenow should be diluted in run mixture tube; BRLs Klenow is 6 units per µl, therefore, should be diluted 1 to 6 for five samples plus one extra.
4. Take 3 µl of Klenow; DTT; ^{35}S dATP reaction mix and add it to the microfuge 1.5 ml tube containing the DNA; mix by pipetting in the out of tip; change tip and then aliquot 3 µl into each appropriate 400 µl N mix tubes (A,C,G,T). Do this for each sample then place in microfuge, spin and incubate rxn for 20 min. at 30°C. After 15 min into 20 min rxn start placing 1 µl of 0.5mM dATP chase into each 400 µl Eppendorf tube. At 20 min remove tubes from bath-spin, place back in bath for an additional 15 min in water bath.
5. Terminate reactions at the end of 15 min 2nd incubation by pipetting 10 µl of formamide-dye mix into each tube. Spin down and place directly on ice or place at -20°C for further use.
6. If reactions are to be run immediately; first prerun gel (15 min minimum) at 1800 V. Near the end of prerun place tube in 90°C and bath for 3 min. place back on ice then load 3-4 µl of each sample. Place samples in -20°C for further use.

iv) Polyacrylamide gel electrophoresis

Reagents

1. 40% acrylamide stock.

Acrylamide	38 g
Bis (NN'-methylene bis acrylamide)	2 g
DDW to	100 ml

To this add 5 g of Amberlite MB1. Stir gently for 30 min. Filter through a sintered glass filter to remove resin. Remove all the fine particles with a Millipore filter and store in dark at 4°C.

2. 10 xTBE buffer

Tris base	108 g
Boric acid	5.5 g
Na_2EDTA, $2H_2O$	9.3 g

Make upto 1 litre with sterile distilled

Protocol

1. Wash the plates well with detergent and water, then wash with ethanol, rinse with distilled water and dry with tissue paper.
2. Siliconise the notched plates. To siliconise the plate, use 2% dimethyl dichlorosilane in 1, 1'1-trichloroethane (BDH repelcote) in a fume cup board. Wear gloves and handle with care. Wipe the plate with tissues and leave to dry (10 min.). Wash with distilled water and wipe dry. Prepare for gel preparation by putting spacer and vinyl tape to prevent leakage.
3. High resolution gels used for dideoxy sequencing are usually 8% or 6%, Polyacrylamide gels containing 8M urea and tris borate buffer.Acrylamide and bisacrylamide are neurotoxins. Avoid inhalation and skin contact. Wear gloves, when handling these materials and when pouring a gel. A typical 6% gel solution (50 ml) is made as follows:

Urea	21 g
Acrylamide (40% soln.)	7.5 ml
10 X TBE buffer	5.0 ml
St H_2O to	50 ml

4. Warm to 37°C to dissolve urea. Add 300 µl freshly made 10% ammonium persulphate solution (APS). Add 50 µl TEMED. Pour gel using a 50 ml syringe (rinse before use).

a) Pouring gels

1. To pour the gel, tilt the plates at an angle of 30°C from the horizontal and slope slightly so that the acrylamide solution can be poured down the side of the plates to slowly fill from the bottom without forming bubbles.
2. Insert comb and clamp plates together using fold back spring slips. Leave to set for at least one hour.
3. Remove vinyl tape. Carefully take out the comb and clamp the gel onto the gel apparatus.
4. Immediately fill the buffer tanks and wash out the wells thoroughly with running buffer (1 x TBE) to remove unpolymerised acrylamide. Just before

loading flush out the sample wells again with running buffer. This is important since urea leaches into the gel slots, preventing the sample from forming a precise layer. Pre-electrophorese for about 3 hrs.

b) Loading the gel

1. Heat the sample at 90-95^{0}C for atleast 30 min. Load onto the gel immediately (Do not cool). As the formamide concentration of the sample drops, the DNA may start to renature. Therefore, it is necessary to load all samples quickly (within 10 min) to avoid obscuring the upper part of the gel with renatured DNA.

c) Running the gel

1. Run the gel in 1 TBE buffer at 15-20mA. The voltage initially will be 1.5KV but as the gel heats, it will drop to 1.2KV. The gel must be hot enough to keep the DNA denatured. A gel temperature of 50-60^{0}C gives good results. This high temperature helps to maintain the oligonucleotides in single stranded denatured state. The use of thermostating plate will give a uniform temperature.
2. For reading the maximum length of sequence two separate loadings on 40cm long gels will be necessary. Run one loading for 1½ hrs and the other for 2 hrs or until the bromophenol dye is just running off the bottom of the gel. Adequate safety precautions should be taken when operating at high voltage.
3. After the run, lift the smaller glass plate, slowly by inserting a spatula in between the plates. Transfer the whole gel onto a Whatman filter paper. Wrap it in saran wrap. Place the gel in the gel drier apparatus for drying and dry it at 60^{0}C for 1 hr. Later dry the gel under vacuum for overnight and use for autoradiography.

v) Autoradiograpy

Reagents

i) X-ray film developer

Water at 50^{0}C	500 ml
Metol	1 g
Sodium sulphite	75 g
Hydroquinone	9 g
Sodium carbonate monohydrate	25 g
Potassium bromide	5 g

Dissolve in order and make up to 1 litre with cold water

ii) Fixer

Sodium thiosulphate	80 g
Potassium metabisulphite	5 g

Dissolve in order in 250 ml and filter

iii) Stop bath 2% acetic acid

iv) X – ray film

v) Cassette with intensifying screen

Protocol

1. Take the dried gel attached to Whatman 3 MM paper. Remove the Saran Wrap.
2. Place pieces of tape marked with radioactive ink at several locations around the edge of dried gel.

   (Radioactive ink is made by mixing a small amount of ^{32}P with a water proof black ink. It is convenient to make the ink in three grades.

Very hot	=	2000 cps on a minimonitor
Hot	=	500-2000 cps on a minimonitor
Cool	=	50-500 cps on a minimonitor

3. When the ink is dry, wrap the sample and backing sheet in Saran Wrap.
4. In a dark room, place the sample in a X-ray film holder and cover it with a X-ray film (use the screen if required. The sensitivity of the film may be increased if intensifying screen is used).
5. Close the casette and wrap it in black bags.
6. Keep at -70^{0}C for several hours to a few days (this depends on the radioactivity on the gel).
7. Remove the film after exposing and develop it in developer for 5 min in dark.
8. Wash the film in stop bath for 1min and fix the film in fixer for 5 min in dark.
9. Wash under running water for 30 min.
10. Dry the film at room temperature and read the sequence by placing the autoradiogram on the light box or read DNA sequence and analyse by having DNASIS in the IBM computer. DOS version 3.0.

B. Maxam Gilbert Method

The chemical method of DNA sequencing was developed by Maxam and Gilbert at Harvard University. These workers made use of chemical reagents to obtain base specific cleavage of DNA. Even though, Dideoxy method of DNA sequencing developed by Fred Sanger at Cambridge University in England is the favoured method, the chemical method, still has use in certain situations. If DNA has some secondary structure, or high GC rich regions then chemical method may be useful.

Chemical method begins by generating a restriction fragment assymetric which is labeled with ^{32}P at one end. The DNA can be either single stranded or double stranded. The base specific cleavage is then carried out with chemical reagents in a series of steps. First, bases in DNA are modified. For example, dimethyl sulphate methylates guanine at the N7 position. The modified base is then removed from the sugar phosphate backbone of DNA. Piperidine cleaves the strand at the sugar residue lacking the base.

In a limited reaction with one end labeled DNA, base specific reagents generate a set of nested end labelled fragments of varying length. The reaction products are analysed by running four samples side by side on a 8% sequencing gel.

i) Labeling of DNA (Klenow fill in)

Protocol

1. Digest 1 to 3 μg of DNA with appropriate restriction enzyme in a total volume of 30 μl of medium or high salt buffer.
2. Add to each tube 2 μl of each of dCTP, dGTP, TTP and dATP. One of the nucleotide should be labelled with ^{32}P and be in complementary sequence of overhang of restriction site
3. Add 1 to 3 μl of Klenow enzyme.
4. Incubate at RT for 30-60 min.
5. After incubation add 2 μl of 10 mM of cold deoxynucleotide which was added as labelled.
6. After 2 min stop the reaction by adding 2 μl of 0.5M EDTA pH 8.0.
7. Phenol extract once. Add 300 μl of 0.3M NaCI or sodium acetate, vortex and centrifuge and precipitate with ethanol. Wash with ethanol, dry under vacuum for 10 min. Resuspend in 30 μl of reaction buffer and then digest with appropriate restriction enzyme. Add proper dye and separate the labelled fragment on 5% polyacrylamide gel.
8. Locate fragment by autoradiography. Cut the gel fragment out under UV light.
9. Add 500 μl of 0.5M ammonium acetate, 1mM EDTA to extract the DNA from gel slice by leaving it overnight at 37^0C. Filter the gel extract through

glass wool which has been plugged in 1 ml Eppendorf pipette tip. Put the pipette tip with glass wool in an Eppendorf tube.

10. Centrifuge the tube containing labeled gel slice for 10-30 min from step 9 and apply the supernatant with 1 ml Eppendorf pipette to the top of glass wool-plugged tip.
11. Wash the labeled gel fragment with 200 µl of 0.5M ammonium acetate and 1 mM EDTA, vortex 1 min., centrifuge for 10 min and repeat the filtration process.
12. Precipitate the hot DNA with ethano and wash. Dry under vacuum for 10-30 min. Suspend DNA in 35 µl of double distilled water and store at -20⁰C

ii) Sequencing reactions

Reagents

i) DMS Buffer : 50 mM Sodium cacodylate, pH 8.0

ii) HZ Stop : 0.3M Sodium Acetate, pH 7.0, 1 mM EDTA

iii) 0.3M Sodium acetate pH 7.0

iv) DMS Stop : 1.5M sodium acetate, pH 7.0, 1M 2mercaptoethanol

v) 10% Formic acid

vi) DMS : Dimethyl sulfate

vii) HZ : Hydrazine

viii) 100% ethanol

ix) Carrier DNA : 1mg/ml DNA in DDW

x) tRNA : 10mg/ml tRNA in DDW (yeast tRNA)

xi) Labelled DNA in DDW

Protocol

1 Set up :

i) For each DNA fragment (labelled) to be sequenced, mark 4 reaction tubes but do no add the reactive chemicals

G	AG	CT	C
200 µl DMS buffer	10 µl ddH_2O	10 µl ddH_2O	10 µl 5M NaCl

2. 1 µl carrier DNA (any plasmid, 1 mg/ml)

5 µl Hot DNA	10 µl Hot DNA	10 µl Hot DNA	05 µl Hot DNA

The tubes should be capped and kept at RT.

ii) Mark up 4 "STOP" tubes

(50 μl DMS STOP)	(200 μl 0.3 M NaAcetate)	(200 μl HZ STOP)	(HZ STOP)	1 μl tRNA (10 mg/ml)

3. DNA Modification Reactions

   i) Add 3 μl 10% Formic acid to AG reaction (15 min, 37ºC). Start 15 min timer

   ii) Add 30 μl HZ to CT (9 min, 16-20ºC)

   iii) Add 30 μl HZ to C (11 min, 16-20ºC)

   iv) Add 1.5 μl DMS to G (5 min, 16-20ºC)

4. While reactions are in progress, add 100% ethanol to the STOP tubes so that they are about 1-1.2 ml full.

5. Stop each reaction by pipetting the contents of the STOP tube into the reaction tube (use the same pipette for all of them), then quickly invert, shake vigorously (very important) and place in a dry ice ethanol bath for 6-7 min (do not leave it there for more than 10 min; at this stage, it is important to work fast because the tubes still contain reactive chemicals).

6. Centrifuge at 4ºC for 7 min, discard supernatant, aspirate the rest of the liquid and then add 1 ml ethanol (100%). Vortex and immediately centrifuge at room temperature for 3 min, then discard the supernatant, aspirate the rest of the liquid (important) and dry in vacum for 10 min.

To the pellets in the bottom of the Eppendorf tubes, add 200 μl of ice cold 3M sodium acetate. You may tip the tube to x the solution with the DNA pellet or you may simply add ice cold 95% EtOH. Invert the tubes several times to mix and place them in the dry ice EtOH bath for 7 min. Centrifuge for 5 min and decant the tubes as before.

Ethanol wash

This is the last step to clean up any salt that may be present in the DNA sample that will create problems when running gels. Thus, it is important to mix the ethanol and to remove as much salt as possible.

To the DNA pellet add 500 μl of 95% EtOH, tip the tube or shake it gently from side to side or even vortex it gently to mix the ethanol around the tube. Freeze the tube in dry ice EtOH for 2-3 min and then spin at room temperature for 2 min. When the centrifuge has stopped, remove as much of the ethanol as you can with a glass micropippete connected to a vacuum line. Holding the tube at 45 degree angle with the pellet at the bottom of the tube, you should be able, by sucking from the top of the solution, to remove all the liquid and be left with a clean DNA pellet. Place all the tubes in a vacuum desicator for a few minutes and prepare for the next reaction.

iii) Piperidine reaction

The piperidine reaction is used to cleave the DNA fragments at the bases where they were modified by the base specific reactions as above. Piperidine reaction is carried out. Heat 100 µl of 10% piperidine in ddH_2O at 80°C for 30 min. Each tube with a dry or fairly dry DNA pellet in the bottom is placed, with the piperidine already added in a 90°C temp block for approximately 30 seconds with the cap open. This allows the tube to heat up & the air above the sample to expand and not create increased pressure inside the tube that may blow the cap off and spill your sample all over. At the end of 30 min or 28 or 25, place the tubes in an Eppendorf centrifuge and spin them for 10-20 seconds to collect all the condensation from the cap into the bottom. While the tubes are still in the centrifuge, use a needle (20 gauges) to pock 2-3 holes in the top of the tubes. Now freeze the tubes in a dry ice EtOH bath and then place them in the lyophilizer until the tubes are completely dry. When the tubes are dry, remove them and resuspend the DNA in whatever amount formamide based loading dye you want.

iv) Polyacrylomide gel electrophoresis and autoradiography

Apply 2 µl per lane on an 8% polyacrylamide gel and to a 20% gel if sequence of the 50 base pairs near the end is required. Prepare sequencing gels as described earlier and run under similar conditions. Save the rest for further use.

When the dye has run 2/3rd of the length of the gel, load 2 µl samples second time. Run the gel till dye is one inch above the front, stop, and proceed for auto-radiography as described in Sanger's method.

Suggested reading

Maxam A. and Gilbert W. (1980). Sequencing end labeled DNA with base specific DNA. In Methods in enzymology vol 65, Grossman L and Moldave K eds. Academic Press, UK.

Sanger F., Nickel S. and Coulson A.R. (1977). DNA sequencing with chain terminating inhibitors. Proc. Natl. Acad. Sci. 74: 5463-5467.

Sanger F. (1981). Determination of nucleotide sequences in DNA. Science 214: 1205-1210.

16

Isolation and *in vitro* Culture of Protoplasts

Protoplasts are cells which have had their cell wall removed, usually by digestion with enzymes. Cellulase enzymes digest the cellulose in plant cell walls while pectinase enzymes break down the pectin holding cells together. Once the cell wall has been removed the resulting protoplast is spherical in shape. Digestion is usually carried out after incubation in an osmoticum (a solution of higher concentration than the cell contents which causes the cells to plasmolyse). This makes the cell walls easier to digest. Debris is filtered and/or centrifuged out of the suspension and the protoplasts are then centrifuged to form a pellet. On resuspension the protoplasts can be cultured on media which induce cell division and differentiation. A large number of plants can be regenerated from a single experiment e.g. a gram of potato leaf tissue can produce more than a million protoplasts. Protoplasts can be isolated from a range of plant tissues: leaves, stems, roots, flowers, anthers and even pollen. The isolation and culture media used vary with the species and with the tissue from which the protoplasts were isolated. Protoplasts are used in a number of ways for research and for plant improvement. The can be treated in a variety of ways (electroporation, incubation with bacteria, heat shock, high pH treatment) to induce them to take up DNA. The protoplasts can then be cultured and plants regenerated. In this way genetically engineered plants can be produced more easily than is possible using intact cells/plants.

Plants from distantly related or unrelated species are unable to reproduce sexually as their genomes/modes of reproduction etc. are incompatible. Protoplasts from unrelated species can be fused to produce plants combining desirable characteristics such as disease resistance, good flavour and cold tolerance. Fusion is carried out by application of an electric current or by treatment with chemicals such as Polyethylene Glycol (PEG). Fusion products can be selected for on media containing antibiotics or herbicides. These can then be induced to form shoots and roots and hybrid plants can be tested for desirable characteristics.

Media and Solutions

Murashige and Skoog medium (MS-medium)

i)	Major salts	Mg/l
	KNO_3	1900
	NH_4NO_3	1650
	KH_2PO_4	170
	$MgSO_4.7H_2O$	370
	$CaCl_2.\ 2H_2O$	440
ii)	Minor salts	Mg/l
	$FeSO_4.7H_2O$	27.8
	$Na_2EDTA.2H_2O$	37.3
	H_3BO_3	6.2
	$MnSO_4.\ 4H_2O$	22.3
	$ZnSO_4.7H_2O$	8.6
	$CoCl_2.6\ H_2O$	0.025
	$CuSO_4.5\ H_2O$	0.025
	$Na_2MoO_4.2\ H_2O$	0.25
	KI	0.83
	Meso-inositol	100
	Nicotinic acid	0.5
	Pyridoxine HCl	0.5
	Thiamine HCl	0.1
	Glycine	2.0
	Sucrose	30,000
	Agar	8,000 (pH 5.8)
iii)	Kao's medium for protoplast culture	Mg/l
	NH_4No_3	600
	KNO_3	1900
	$CaCl_2.2\ H_2O$	600
	$MgSO_4.7H_2O$	300
	KH_2PO_4	170
	KCI	300

Sequestrene 330 Fe	28
Glucose	72,000
Sucrose	125
Fructose	125
Ribose	125
Xylose	125
Sodium pyruvate	5
Citric acid	10
Inositol	100
Thiamine HCl	10
Ascorbic acid	1
Nictotinamide	1
KI	0.75
H_3BO_4	3.0
$MnSO_4.H_2O$	10.0
$ZnSO_4.7H_2O$	2.0
$Na_2M0O_4.2H_2O$	0.25
$CuSO_4.5H_2O$	0.025
$CoCl_2.6H_2O$	0.025
Mannose	125
Rhamnose	125
Cellobiose	125
Sorbitol	125
Mannitol	125
Malic acid	10
Fumaric acid	10
Folic acid	0.2
Riboflavin	0.1
p-amino benzoic acid	0.01
Vit. B_{12}	0.01
Vit. A	0.005
Vit. D_3	0.005
Biotin	0.005

Coconut water	10 ml
Pyridoxine HCl	1
Calcium pantothenate	0.5
Choline chloride	0.5
Casein hydrolysate	100
2, 4-D	1.0
BAP	0.5
NAA	0.1

(pH 5.7)

(Filter sterilize)

iv) Preplasmolysis solution

MS major salts (half strength)

MS minor salts (half strength)

$CaCl_2 . 2H_2O$	0.5%
Mannitol	7.2%

pH 5.8

(Sterilization by autoclaving)

v) Enzyme solution

MS major salts (full strength)

MS minor salts (full strength)

Mannitol	7.2%
Cellulose R-10 (Onozuka)	1%
Pectolyase Y-23	0.025% (0.25% Macerozyme R10 for leaf protoplasts)
pH	5.6

(Filter sterilize)

vi) 21 S Soution

sucrose	21%
$CaCl_2$	0.5%
MES	5mM
pH	5.6

(Sterilization by autoclaving)

vii) Protoplast wash solution (W_5)

NaCl	2.2 g
$CaCl_2$	4.6 g
KCl	0.100 g
Glucose	0.250 g
Mannitol	7.5 g

Make upto 250 ml (pH 5.6)

viii) Protoplast dilution medium (K3)

Major	Mg/l
NH_4NO_3	250
KNO_3	2500
$CaCl_2\ H_2O$	900
$MgSO_4.7\ H_2O$	250
NaH_2PO4	150
$(NH_4)_2SO_4$	134
$CaHPO_4.\ H_2O$	50
Micro	
$CoCl_2.6\ H_2O$	0.025
$CuSO_4.5\ H_2O$	0.025
$Na_2MoO_4.2H_2O$	0.25
$FeSO_4.7\ H_2O$	27.8
$Na_2EDTA.2\ H_2O$	36.3
H_3BO_3	3.0
KI	0.75
$MnSO_4.4\ H_2O$	13.2
$ZnSO_4.7\ H_2O$	2.0

ix) Vitamins

Nicotinic acid	1.0
Pyridoxine HCl	1.0
Inositol	100
Thiamine HCl	10.0

x) Sugars

Xylose	250
Sucrose	3.42%

xi) Growth regulators

2,4-D	
BAP	1.0 mg/l

pH 5.8

For plating use K3 medium with 0.4% agarose (Type I, Low EEO Sigma)

Protocol

(All work should be carried out in a laminar flow)

1. Surface sterilize *B.Juncea* seeds by treating with 0.1% $HgCl_2$ for 12-15 min
2. Thoroughly wash treated seeds by rinsing (4-5 times) with sterile distilled water.
3. Transfer 8-10 seeds/tube containing half-strength MS-agar medium. (About 12 tubes will be needed for one isolation experiment).
4. Incubate in dark at $25^0 \pm L2^0C$ for 7 days.
5. Take shoot tips from *in vitro* light grown seedlings and incubate onto half-strength MS-agar medium.

i) Isolation of protoplasts

1. Take 7-day old etiolated hypocotyls of *B.Juncea* on a sterilized petri-dish.
2. With the help of a sharp scalpel cut about 50 hypocotyls into small segments (5 mm) and aseptically add 15 ml of pre-plasmolysis solution containing 0.5% $CaCl_2$ and 7.2% mannitol (pH 5.8). Leave for 1 h. (Similarly repeat for leaf tissue).
3. Remove pre-plasmolysis solution with a sterile Pasteur pipette and add 10-15 ml of enzyme mixtgure. Seal the dish with parafilm
4. Incubate in dark at 25^0C for 16-18 h.
5. Release the protoplasts by gently drawing through Pasteur pipette (8-10 times).
6. Pass the enzyme mixture along with protoplasts and cell debris through a stainless steel mesh (250 nm).

7. Collect the filtrate and put in a 15 ml centrifuge tube and centrifuge at 300 g for 5 min. (Use 100 x g for leaf protoplasts).
8. Take and discard supernatant with the help of a Pasteur pipette.
9. Suspend the pellet in 21S sucrose solution and spin for 10 min. at 300 g.
10. Collect protoplasts that float at the top with a Pasteur pipette..
11. Dilute to about 1:10 with protoplast wash solution. Centrifuge at 300 g (5 min).
12. Discard the supernatant and repeat washing once again.
13. After discarding supernatant, add 1 ml of protoplast culture medium (Kao's medium) to the pellet and suspend the protopiasts.

ii) *In vitro* culture

14. Draw a small aliquot of protoplast suspension and count the density of protoplasts with the help of a haemocytometer.
15. Adjust the density of protoplasts to 105/ml by adding protoplast culture medium.
16. Dispense 2 ml of protoplast suspension per 30 mm petri dish. Seal and incubate in dark at 25^{0}C after 4th day.
17. Transfer to light (16 h), 25^{0}C after 4th day.
18. On 7th, 10th and 13th day, add 0.5 ml of protoplast dilution medium (K3) to each dish. Incubate in light in 25^{0}C.

    (After 1-2 days cell wall formation and cell divisions can be observed. Small cell clusters are visible by 7 days).
19. After 19 days plate cells on agarose medium (K3)

iii) PEG mediated fusion of protoplasts

Polyethylene glycol is a chemical fusogen widely used in somatic fusion of both plant protoplasts and animal cells. Application of a highly concentrated solution of PEG draws out H_2O from protoplasts and brings together the protoplast such that the membranes of protoplasts come into close contact. When PEG solution is diluted, protoplasts enlarge and the membranes of adjacent protoplasts fuse at the point of contact. Thus somatic fusion is accomplished.

Reagents

1. Fusion solution

PEG 8000	20% (w/v)
DMSO	10% (v/v)
Glucose	3.6% (w/v)
$CaCl_2.2H_2O$	0.17% (w/v)
KH_2PO_4	0.0095% (w/v)
Filter sterile	(pH 5.8)

2. Wash solution (W_5)
3. Protoplast culture medium (Kao's medium)

Protocol

1. Isolate protoplasts and suspend in culture medium.
2. Mix equal volume of both the hypocotyl and the leaf mesophyll protoplasts of B. Juncea and B. oleracea in a centrifuge tube.
3. Place two drops (40-50 μl each) of protoplast mixture on a 30 mm sterile disposable dish. Allow protoplasts to settle for 5 min.
4. Gently introduce from one side equal volume of fusion solution to each of the droplets. Leave undisturbed for 10 min.
5. Draw out with a Pasteur pipette most of the solution from each of the droplets and discard. (Take care not to remove all solution, protoplasts will collapse).
6. Add 40 μl of wash solution (W5 medium) to the protoplast droplets. Allow to stand for 2 min.
7. Repeat step 6 two more times. Allow to stand for 10-15 min.
8. With the help of a Pasteur pipette draw out most of the most solution and discard.
9. Add 1.5 ml of protoplast culture medium to each dish. Seal with parafilm.
10. Incubate in dark at 25^{0}C for 3 days. (Follow steps 17-19 of protoplast culture for further culture protocol of fusion products).

iv) Electroporation of tobacco protoplasts

Electroporation is a physical reversible process leading to the formation of transient pores in the cell membrane of the protoplasts which would allow the transfer of nucleic acids and other macro-molecules inside. In general, genetic modification in prokaryotes is referred to as transformation and in eukaryotes, the process is called transfection. Since it is based on electrical discharge, it is more appropriately called electro-transfection. This reversible process of creating pores in cell membrane is mediated by a DC pulse discharged by a capacitor in an electroporator.

Protocol

1. Wash the purified tobacco mesophyll protoplasts with W_5 solution.
2. Wash protoplasts once again in 0.6 M mannitol solotion at pH 7.0-7.5 at 100 x g for 3-5 min.
3. Suspend the pellet in 0.6M mannitol at 106 protoplasts per ml.
4. Dispense 250 µl of protoplast suspension to the pre-sterilized electroporation cuvette with absolute alcohol for atleast 15 min
5. Sterilize electroporation electrodes with absolute alcohol for atleast 15 min
6. Add 5-10 µg DNA (pBl 221 with 35S-GUS)
7. Electroporation settings

Choose mode T	500 V
Resistance R	R 9 (480 ohms)
Capacitance C	100 µF
Cuvette gap	2 mm
Field strength	0.5 kv/cm
Pulse duration	32 msec.

8. After electroporation, keep protoplasts on ice for 10 min. before plating as described earlier.

Suggested reading

Krens F.A., Molendijk, L., Wullema G.J. and Shilperoot, R.A. (1980). *In vitro* - transformation of plant protoplasts with Ti plasmid DNA. Nature, 296: 72-74.

Murashige and Skoog F. (1962). A revised medium for rapid growth and bio assays with tobacco tissue cultures. Physiologia plantarum 15(3): 473-497.

Paszowski J., Shillito R.D., Saul M., Manda K.V., Hohn T., Hohn B. and Putrykus I. (1984). Direct gene transfer in plants. EMBO J. 3: 2717-22.

17

Methods of Genetic Transformation of Plants

Developments in recombinant DNA technology have made it possible to transfer gene from any source across species barrier. In fact, transfer and expression of foreign genes in plant cells is now a routine procedure. Of the several methods tried so far, the system based on Ti plasmid of *Aprobacterium tumefaciens* appears to be most successful for transformation of plant cells. Cauliflower mosaic virus based vectors have not made any impact due to severe limitations on the size of DNA that can be cloned. Replicons of other viruses with broad host range are being examined to develop vectors for plant transformation. Direct uptake of DNA (preferably linear) by protoplasts can be used, but it requires more skill and time. Direct DNA transfer remains the method of choice for plant species not susceptible to *Agrobacterium*. More recently, shooting microprojectiles coated with DNA has been quite successful in introducing DNA directly inside plant tissues especially somatic embryos. This method seems to have a great potential in transformation of plants not susceptible to *Agrobacterium*.

Transformation with *Agrobacterium* may be done by co-cultivation of protoplasts, leaf discs or any other explant with the bacterium. Leaf pieces are most popular perhaps because of convenience and low background in selection methods.

A.tumefaciens has the ability to produce tumors on dicotyledonous plants. A large plasmid (about 200 kbp) called Ti (tumor inducing) plasmid is responsible for this. It transfers and inserts a particular segment of its DNA called T-DNA (10 to 20 kb) in the plant genome. T-DNA has been shown to be stably integrated into plant genome and transmitted through seeds. The tumorigenic genes from T-DNA have been removed in plant vectors so that tumors are no longer formed by such truncated vectors and plants can be regenerated from the transformed tissue. The ends of the T-DNA (border sequences) are very important functionally in the transfer and integration of T-DNA. The right border is very important and capable even when alone (in the absence of left border) to allow transfer DNA. Also present on Ti plasmid is the *Vir* region which can function even *in trans*. The *chv*B and *psc*A are the constitutively expressed chromosomal virulence loci which encode products involved in the binding of *Agrobacterium* to plant cells during the infection process.

Two different vector systems have been developed

1. **Co-integrative vector system**: In this the sequence of interest is inserted into the truncated T-DNA by a single homologous recombination event between similar sequences in the T-DNA and the cloning vector. The *Vir* genes are in cis. *A.tumefaciens*. (pGV3850::1103 neo), contains a cointegrate between pGV3850 and pLGV1103. Such strains are prepared by mobilization of pLGV1103 from *E.coli.* to *Agrobacterium.*
2. **Binary vector system**: In this the sequence of interest is directly cloned between T-DNA border sequences, on specially designed vectors which can replicate both in *E.Coli.* and *Agrobacterium.* The air genes are present *in trans* on a Ti plasmid without tumorigenic functions.

Extensive research work has now led to elucidated of the molecular mechanisms of the transfer of DNA from *Agrobacterium* to plant cells. Based on the knowledge generated *agrobacterium* strains and efficient vectors have been developed. The present protocol describes the methodology for transformation of tobacco using a disarmed *Agrobacterium* strain LBA 4404 containing a binary vector p35S GUSINT. This binary vector contains a kanamycin resistance gene (*npt* II) and a β-*glucuronidase gene (gus*) as a selectable and scorable marker, respectively.

A. *Agrobacterium* mediated Transformation of Tobacco

Reagents

i) Co-cultivation

Agrobacterium strain LBA 4404 (p35SGUSINT)

Aseptically grown Tobacco (*Nicotiana tabacum*) plants

Ethanol	
Kanamycin	(100 mg/ml)
Cefoxtaxime	(250 mg/mi)
Liquid medium	
YEMB	
K_2HPO_4 solution	10 ml
Mannitol	10 g
Yeast extract	1g
$MgSO_4$ (1M)	0.8 ml
K_2HPO_4	5 g

NaCl	1 g
H_2O	100 ml
MSO (MS medium without phytohormones)	
MS 104	1.9 mg/L BAP
MSO	0.1 mg/L NAA
Solid medium	
Yema	1.6 Agar/100 ml YEMB
MS 104(S)	0.7 g Agar in 100 ml MS 104
MSO(S)	0.7 g Agar in 100 ml MSO
Selection medium	

100 mg/L kanamycin + 250 mg/L cefotoxine in MS104 (DS).

Protocol

1. Grow the *Agrobacterium* strain in 5 ml of YEM medium, overnight at 28°C.
2. Tissues from aseptically grown sterile plants are most commonly employed. Tissues from non-aseptically grown plant must be surface sterilized before use.
3. Chop the tissue into small pieces (about 0.5 cm^2 sections for leaf tissue and about 0.5 cm long section for stems and hypocotyls).
4. Place several pieces of plant tissue in a small Petri dish (50 mm). Add 4 ml of MS I04 medium.
5. Add 50 µl of an overnight grown culture of *Agrobacterium.*
6. Seal the Petri plate with parafilm and cocultivate bacteria with plant tissue for 2days at 28°C in the dark.
7. Wash the plant tissue with MSO medium and blot them dry with sterilized filter papers.
8. Incubate the tissue for a few days on a non-selective MS 104 (S) medium containing 200 µg/ml cefotaxime.
9. After 5-7 days, transfer the tissue to selective medium MS 104 containing 100 µg/ml Kanamycin and 200 µg/ml Cefotaxime.

ii) Characterization of transformants

The transformed nature of plant/plant tissue should be confirmed by performing a Southern hybridization analysis by using marker gene DNA as probe and by studying the expression of a reporter enzyme (e.g. β-*glucuronidase*).

a) Southern hybridization analysis

Protocol

1. Isolate high molecular weight DNA from transformed and untransformed plant tissue.Digest the plant DNA with *ECoRI* and electrophorese the uncut and digested DNA on 0.7% agarose gel.
2. Carry out blotting of this gel on a nylon membrane.
3. Preparation of probe: Isolate and purify the plasmid carrying *npt II* gene.
4. For preparing probe nick translate the whole plasmid using radiolabelled dCTP.
5. Carry out hybridization of the blot.
6. Expose the membrane to the film.
7. Develop the autoradiogram.

b) Marker genes assay

Marker gene *[β-Glucuronidase (GUS)]* Assay

1. Extraction Buffer 50 mM Na_2 HPO_4; 10 mM EDTA; 0.1% Triton; 0.1% Sodium Lauryl Sarcosine; 10 mM 2-mercaptoethanol; pH 7.0 adjusted with NaOH.
2. Substrate 4-methyl umbelliferyl p-D glucuronide (PIUG), 0.1 M, in extraction buffer.
3. 0.2M Na_2CO_3.

Protocol

1. Freeze the tissue in liquid nitrogen and grind using a mortar and pestle in extraction buffer (use 2 ml of extraction buffer per g tissue).
2. Centrifuge the extract at 15,000 x g for 15 min.
3. Use the supernatant for *GUS* assay (enzyme extract).
4. Incubate the following mixture at 37°C.

Extraction buffer	960 μl
Substrate	20 μl
Enzyme extract	20 μl

5. Withdraw 200 μl aliquotes from the reaction mixture at zero time and subsequent times (10, 20 and 60 min. etc.) and terminate the reaction by adding 0.8 ml of 0.2 M Na_2CO_3.

6. Fluorescence is then measured with excitation at 365 nm, emission at 455 nm.
7. Alternatively, the fluorescence can be conveniently detected by placing the tubes on a long wave UV light box.The florimeter should be calibrated with freshly prepared standards of 4-methyl umbelliferrone (100 nM-1mM).

iii) Histochemical localization of β-*Glucuronidase*

Solutions

1. Fixation solution 2% Formaldehyde; 50 mM Sodium phosphate (pH 7.0); 0.1% Triton X-100.
2. Washing buffer 50 mM Sodium phosphate (pH 7.0).
3. X-Gluc Assay buffer1 mM X-Gluc, 50 mM Sodium phosphate (pH 7.0), 0.1% Triton X-100.X-Gluc (5 bromo-4-chloro 3-indolyl β-D-glucuronide) is the most commonly employed for substrate for histochemical localization of *gus* in transgenic plants.

Protocol

1. Using a scalpel or a rozor blade cut thin (0.5-1 mm) sections from plant leaves or stems.
2. Place these tissue sections in 200 μl of fixation solution and vacuum infiltrate.
3. Incubate at 30^{0}C for 20 min.
4. Wash the tissue sections thoroughly (5-6 times) in washing buffer.
5. Incubate the sections in X-gluc-assay buffer (100 μl) at 37^{0}C.
6. Rinse the tissue in 70% ethanol for 10 min.
7. Mount the tissue on a microscope slide and view under a microscope.

B. Particle Delivery System (Biolistics Gun) mediated Transformation of Plant Cells

A large number of gene transfer methods are currently available. However, the method of gene transfer using the Particle Delivery System is most versatile and proved successful in introducing foreign genes not only to the nucleus but also to the chloroplasts. The plant cell wall is usually impermeable to foreign DNA. Utilization of the gene gun method has shown better results against the various barriers that hinder the delivery of foreign DNA inside the organelle/ cell/tissue. Gene transfer through a gene gun is part of the biolistic method. During this method DNA or RNA construct adheres to biological inert particles (such as gold or tungsten) to form a DNA/particle complex. The DNA/particle complex is bombarded on a targeted tissue in a vacuum. This bombardment is achieved by accelerating a powerful shot to the targeted tissue. The high density of the DNA/particle complex increases the bombardment speed toward the targeted tissue; whlch may result in effective transformation. Only tried metal particles can be bombarded through a solution containing DNA surrounding the cell. Such a complex once formed is directly transferred to the plant cell. The application to cells and embryos uses tungsten or gold spheres (app. 1 micron diam) on which DNA adsorbed. The beads are accelerated by stream of helium, stable transformation occurs through illegimate and random recombination of the few (1-5 copies/genome) surviving DNA molecules/cell. The protocol described blow has been standardized for tobacco transformation using tungsten particles.

i) Preparation of microprojecties

1. Use 50 mg of tungsten powder (BioRad Tungsten M-17) for 100 shots
2. Add 1 ml of freshly-operated absolute ethanol
3. Centrifuge for 10 sec in microfuge
4. Replace with 1 ml absolute ethanol and centrifuge for 10 sec. Can be frozen over night at -20°C
5. Remove ethanol. Add 1 ml water, resuspend tungsten and recentrifuge. Repeat 3 times.
6. Finally resuspend tungsten powder in 1 ml water.
7. Sample preparation for 6 shots: 50 μl Tungsten solution
10 μl DNA in TE buffer (atleast 1 μg/μl)
50 μl 2.5 M $CaCl_2$
20 μl 0.1 M Spermidine (free base)
8. Add 200 μl ethanol
9. Centrifuge for 10 sec in a microfuge
10. Remove supernatant. Add 200 μl ethanol and resuspend extremely well, making sure all clumps are dispersed. Repeat 3 times.
11. Resuspend final pellet in 30 μl ethanol
12. Put 5 μl onto each disk.

ii) Preparation of leaf samples for shooting

Protocol

Each shot requires a target leaf, and for one construct, you need to cut 20-30 leaves. For cutting, use sterile forceps and scalpels. Each cut leaf is placed abaxial side up on top of a sterile filter disc (Whatman #4 circles, on an MS plate. Try to centre the leaf in the plate, to maximize the surface receiving the shot. The leaf surface should touch the MS medium as much as possible. Therefore if the leaf is slightly curled up, nick it to allow it to lie flat.

Leaves should be cut a day before transformation. A stack of thin plates containing cut leaves is bagged in plastic and left in the culture room overnight.

iii) Shooting leaf samples

Protocol

For each shot one 1100 p.s.i. rupture disc (the small dark brown discs), one flying disc (the large orange discs), and one stopping screen (the wire mesh screens are required). All these are supplied in the Board shooting kit (500 standard pressure kit Cat No. 165-2283). Before shooting sterilze the discs and stopping screens by placing them for 5-10 min in a petri dish containing absolute ethanol. The discs and screens are then dried by standing them up along the side of a sterile petri dish and allowing them to air- dry under sterile conditions in the tissue-culture hood.

While sterilizng the disposable shooting components, the gun parts should also be surface sterllized by wiping with absolute ethanol. Three components can be removed from the gun vacuum chamber for surface sterilization. These are: the holder for the rupture discs, the stage and holder assembly for the flying discs, and the platform for the leaf samples. After removing these and wiping with ethanol, the interior of the vacuum chamber and door should be wiped with absolute ethanol. Finally the five metal ring holders for the flying discs should be surface sterilized.

1. Turn the gun on shooting.

   Turn the pump on.

   Turn the gas cylinder on. First, turn the screw bar counter clockwise until it feels loose. Then turn the knob on top of the cylinder counter clockwise (the pressure valve on the right should read at least 1300 p.s.i.). Then turn the screw bar slowly clockwise until the pressure valve on the left reads 1100 p.s.i.

2. Place the five flying disc holders in a petri plate and into each one, place a flying disc (the large orange ones). Press down the discs with the red plastic cap so that they fit tightly into the holders.
3. Onto the surface of each disc, pipette 5 µl, of DNA-coated tungsten.
4. Do all of the following steps in the same order for each shot
   i) Place a rupture disc into the rupture disc holder, using forceps. Make sure the disc is properly seated in the holder by tapping the holder a few times gently. Screw the holder into the vaccum chamber. It must be screwed tightly.
   ii) Place a stopping screen onto the diaphragm in the flying disc holder assembly. On top of this, place upside down a metal ring containing a flying disc with DNA on it. Secure all the components by screwing on the assembly ring. Place the stage in the second slot from the top in the vacuum chamber
   iii) Take a petri plate with a leaf sample and place it on the sample plateform (lid off). The sample platform should be positioned in the chamber in the fifth slot from the top (second from the bottom).
   iv) Press the vacuum button (middle red button, up position) to begin pumping air out of the shooting chamber. Allow the vacuum pressure to reach at least 25 inches of Hg. Then press the shooting button continuously (red button on the right, up position). When the pressure reaches 1100 p.s.i., the rupture disc inside the gun should break (you will hear a pop, and see the petri plate jump). Release the shooting button, and remove the vacuum by setting the vacuum button to vent (middle position).
   v) Take the petri plate out of the chamber. Repeat until all samples have been processed.
   iv) Histochemical localization of *β-Glucuronidase*

Leaves can be shot with a gene engineered for nuclear expression from the ^{35}S promoter. Transient expression of this construct can be monitored in shot leaves, 24 to 48 hrs after shooting by staining the shoot leaves for *gus* activity according to the protocol described earlier. If the tungsten coating and shooting conditions were optimum, one should see ~100 blue spots on the shot leaf surface.

Suggested reading

Alam M.F., Datta K., Abrigo E.,Vasquez A., senadhana D. and Datt, S.K. (1998). Production of deep transgenic rice plants expressing a synthetic *Bacillus thuringiensis* 1A(b)gene with enhanced resistance to yellow stem borer. Plant Science 135: 25-30.

Christou, P. (1996). Particle Bombardment for Genetic Engineering of Plants. Academic Press, UK.

Hooykaas P.J.J. and Schilperoot R.A. (1992). *Agrobcterium* and plant genetic engineering. Plant Molecular Biology 19: 15-38.

Lawrence P.K. and Koundal K.R. (2001). *Agrobacterium tumefaciens*-mediated transformation of pigeonpea (*Cajanus cajan* L. Millsp.) and molecular analysis of regenerated plants. Current Science 80(1): 101–105.

Potrykus I. (1990). Gene transfer to plants: assessment and perspective. Physiologia Plantarum 79: 129-134.

[illegible]

[illegible]

[illegible]

18

Restriction Fragment Length Polymorphism (RFLP) and Random Amplified Polymorphic DNA (RAPD) Analysis

The genetic information which makes up the genes of higher plants is stored in DNA sequences of the nuclear chromosomes and the organelle genome. There is such an enormous amount of DNA in higher plant cell, that no two organisms are likely to be identical in DNA base sequence. When nuclear DNA is restricted by restriction enzyme, the fragments produced will thus be specific for each target DNA/restriction enzyme combination. Such differences in fragment size arising from restriction enzyme digestion of nuclear, organelle or total DNA are called RFLP denoting Restriction Fragment Length Polymorphism and can be used as a direct measure of genetic variability. In recent years gene cloning technology has led to the fundamental advances in many areas of biology. One of the most exciting of these is the use of cloned fragments of chromosomal DNA as general markers, usually termed as 'RFLP mapping. In this technique, which depends on natural variation in DNA base sequence, DNA is digested with retraction enzyme. Homologous restriction fragments of DNA which differs in size or length can be used as genetic markers to follow chromosome fragments through genetic crosses. One of the ways to reach this variation is by restriction enzymes. A large piece of DNA is reduced to a series of smaller fragments of defined size by digestion with restriction enzymes. The number of fragments produced and size of each fragment will reflect the distribution of restriction sites in the DNA. The fragment produced will thus be specific for each target DNA/ restriction enzyme combination and can be used as a 'finger print' specific for a given target DNA or for the organism containing that DNA. If digested DNA is subjected to gel electrophoresis and stained with ethidium bromide, no distinct fragments can be visualized. So many fragments are produced that the DNA appears to run as a continuous smear. However, individual restriction fragments are still well resolved in gel and RFLPs are still present between DNAs from different organisms.

More complex techniques have to be used to resolve them. These techniques involve the use of cloned DNA probes and DNA hybridization. Using the high specificity of DNA-DNA hybridization, such probes can detect individual restriction fragments in the complex mixture of fragments of nuclear DNA present in the restriction digest. But a supply of the individual fragments in pure form is needed and it would be very difficult to directly isolate the individual restriction fragment. Now, this is easily done by constructing genomic library of smaller fragments (2-5 kb) size in plasmid DNA. The bacterium replicates the plasmid as it grows and divides. By growing a culture of these transformed bacteria and then isolating the plasmid, one obtains a large supply of a single plant DNA restriction fragment which is suitable for the use as a hybridization probe to detect RFLPs in the nuclear DNA of the organism.

A. RFLP- Restriction Fragment Length Polymorphism

i) Isolation and purification of high molecular weight plant DNA

Reagents

DNA Extraction Buffer

- 100 mM Tris-Cl (pH 8.0); 20 mM EDTA (pH 8.0); 1.4 M NaCl;
- 2.0% (W/V) cTAB; 0.2% 2-Mercaptoethanol
- Chloroform-Isoamylalcohol (24 : 1)
- TE Buffer : 10 mM Tris (pH 8.0); 10 mM EDTA

a) Isolation of DNA

Protocol

1. Weigh 15 g of leaf tissue (etiolated), freeze quickly in liquid nitrogen and grind to a fine powder in a 3" mortar and pestle. It is important not to let the tissue thaw once frozen until buffer is added.
2. Add 75 ml of DNA extraction buffer. For maximum yield, the cells are further broken by grinding the mixture at a low setting (about 3) with a polytron. However, this step is optional.
3. Transfer to a 250 ml flask, add 5 ml of 20% SDS, mix thoroughly by shaking vigorously for 20 min, then incubate tubes at 65^{0}C for 10 min.
4. Add 50 ml of 5 M potassium acetate. Shake flask vigorously and incubate at 0^{0}C for 20 min (keep on ice). Most proteins and polysaccharides are removed as a complex with the insoluble potassium dodecyl sulfate precipitate.
5. Spin tubes at 25,000 g for 20 min. Pour supernatant through a Mira cloth filter into a flask containing 78 ml isopropanol. Mix and keep tubes at -20^{0}C for 30 min.

6. Pellet DNA at 20,000 g for 15 min. Gently pour off supernatant and dry pellets by inverting the tubes on paper towels for 10 min.
7. Redissolve DNA pellets in 3.5 ml buffer of 50 mM Tris, 10 mM EDTA (pH 8.0). Transfer the solution to an Eppendorf tube. Spin the tube in Eppendorf centrifuge for 10 min to remove insoluble debris.
8. Transfer the supernatant to a new Eppendorf tube and add 375 µl of 3 N sodium acetate and 2.5 ml isopropanol. Mix well and centrifuge for 1 min in a Eppendorf centrifuge to pellet DNA. Wash pellet with 80% ethanol, dry using nitrogen gas stream or under vacuum and redissolve in 500 µl of TE buffer.Precipitation with 3M sodium acetate using relatively small amounts of isopropanol (about 0.6 vol.) has been reported to separate high molecular weight DNA from polysaccharides. Sodium acetate also yields a tight fibrous precipitate that is easily washed and dried.

b) Purification of DNA by ultracentrifugation

1. Add 4.5 g of CsCI and 200 µl Ethidium bromide solution.
2. Pellet the precipJate, if any, and transfer the supernatant using syringe (5 ml) and needle (18 gauges) to polyallomer quick seal centrifuge tubes.
3. Heat-seal the tube after proper balancing.
4. Spin the tube at 60,000 rpm in Ti 65 rotor for 18 hrs at 20^{0}C.
5. Visualize DNA band using UV torch and take out the DNA band with the help of syringe (2 ml) and needle (18 gauge) and transfer to Eppendorf tube.
6. Add equal volume of butanol saturated with TE to remove Ethidium bromide.
7. Mix by gentle inversion and spin for 2 min.
8. Carefully remove the upper phase and discard.
9. Repeat steps 6-8 till the lower phase becomes colourless.
10. Transfer the lower phase to a 30 ml corex tube.
11. Add two volumes of sterile distilled water and mix.
12. Add two volumes of cold ethanol and mix by quick gentle inversion.
13. Spool out DNA, wash for 10 min in 70% ethanol and 10 mM ammonium acetate.
14. Dry, dissolve DNA in 500 µl of TE and estimate concentration Spectrophotometrically.
15. Store DNA at -20^{0}C.

ii) Restriction of DNA and agarose gel electrophoresis

Reagents

- 10 X TBE buffer
- 0.89 M Tris; 0.89 M boric acid; 0.02 M EDTA Ethidium bromide (10 mg/ml)
- 10 x gel loading buffer
- 0.25% Bromophenol blue. 0.25% Xylene cyanol; 25% Ficoll (type 400) in water;
- 5M Ammonium acetate
- 0.5 M EDTA, pH 8.0
- BSA (1 mg/ml)

Protocol

1. Add the components of the reaction mixture to an Eppendorf tube in the following order.

Water	130 µl
10 x TBE buffer	20 µl
BSA	20 µl
DNA	20 µl (10 µg DNA)
Restriction enzyme (5 units/µg DNA)	10 µl (50 units)

2. Mix by gentle tapping and incubate at 37°C overnight.
3. Add 8 µl of 0.5 M EDTA to stop the reaction.
4. Add 2/3 volume of 5M ammonium acetate and mix.
5. Precipitate DNA by adding two volumes of ethanol.
6. Pellet DNA and wash in 70% ethanol.
7. Dry and dissolve in 30-40 µl of TE.
8. Carryout agarose gel electrophoresis as described earlier.

iii) Vacuum blotting of DNA from agarose gel

Reagents

1. Depurination solution 0.25 N HCI
2. Denaturation solution 0.4 N NaOH - 0.6 M NaCI
3. Neutralisation solution 1.5 M NaCl - 0.5 M Tris-HCI, pH 7.5

4. Transfer solution 10 x SSC (1.5 M NaCI, 0.15 M sodium citrate)
5. 0.4 N NaOH
6. 0.2 M Tris-HCI, pH 7.5-2 x SSC

Protocol

1. Incubate gel in Depurination solution for 10 min at room temperature.
2. Wash gel in sterile distilled water and incubate in denaturation solution for 30 min at room temperature.
3. Remove gel, wash in distilled water and incubate in neutralisation solution for 30 min at room temperature.
4. Cut gene screen plus nylon membranes accroding to the size of the gel and place a mark on its concave side.
5. Carefully place the membrane onto the surface of deionised water and allow membrane to wet by capillary action.
6. Remove the gene screen plus and lay onto 10 x SSC Let it soak for 15 min.
7. Cut a plastic mask, and make an opening on window for the gel such that the gel is slightly bigger than the window.
8. Connect the LKB vacuum blotting unit with the vacuum pump and in-line liquid trap.
9. Place the porous support screen with the shiny side up, on the inner rim of the base unit.
10. Place the plastic mask with the window on the support screen.
11. Position the pre-treated nylon membrane carefully within the window. Avoid trapping air bubbles between the membrane and the porous support screen.
12. Slide the gel from the casting tray onto the membrane to fill the window. Avoid trapping air bubbles.
13. Fit the top frame by evenly tightening the four locking nuts. Switch on the vacuum pump to immobilise the gel.
14. Mark the well position with a pencil and seal the wells with molten agarose.
15. Pour on the transfer solution to cover the gel and leave for one hr.
16. Remove the transfer solution, turn off the pump and take the gel out.
17. Remove the membrane, immerse in an excess of 0.4 N NaOH for one min.
18. Take out the immerse the membrane in an excess of 0.2 M Tris-HCI, pH 7.5, 2 x SSC.
19. Allow membrane to dry at room temperature.

20. Bake the membrane at 80⁰C for 2 hrs. under vacuum.
21. Store the membrane under vacuum at room temperature.

iv) Large scale isolation of plasmid DNA

The starting point for isolation characterization and genetic mapping of clone's will be library of nuclear genomic DNA as described earlier. The incumbinent plasmid isolated from the overnight grown transformed *E.coli* culture and used for making prob for Southern blotting.

Reagents

Per litre LB medium

Bacto tryptone 10g; Bacto yeast extract 5 g; NaCl 10 g; Adjust pH to 7.5 with NaOH, Autoclave.

Solution I

50 mM Glucose; 25 mM Tris Cl (pH 8.0); 10 mM EDTA; Lysozyme solution 10 mg (1 ml) in solution I

Solution II

0.2 N NaOH; 1% SDS

Solution III

3 M Sodium acetate (pH 4.8)

TE Buffer

10 mM Tris Cl pH 8.0; 1 mM EDTA

Isopropanol

Ethidium bromide (10 mg/ml in H_2O

Ethanol

70% ethanol

n-Butanol saturated with TE

Protocol

1. Pick single colony of the desired strain of *E. coli* into 5 ml of LB containing selective antibiotic and grow it overnight at 37⁰C.
2. Inoculate 50 μl of the overnight culture into 500 ml LB containing selective antibiotic and grow overnight at 37⁰C.

3. Harvest the bacterial cells by centrifugation at 4000 g for 10 min at 4ºC.
4. Resuspend the bacterial cells in 10 ml of cold solution I.
5. Add 20 ml of solution II. Mix by inverting the tube. Keep at 40ºC for 15 min.
6. Add 15 ml of 3 M sodium acetate. Mix by inverting the tube. Keep at -20ºC for 90 min.
7. Filter through Mira cloth and centrifuge the filtrate at 5,000 g for 15 min.
8. Add 180 ml isopropanol. Mix well and centrifuge at 5,000 g for 10 min.
9. Redissolve the pellet in 6 ml TE.
10. Add 800 µl of EtBr and 1.1 g of CsCl per ml solution.
11. Completely fill Beckman's quick seal tube and heat seal.
12. Load the tube on Ti-65 type rotor.
13. Centrifuge at 60,000 rpm at 20ºC overnight.
14. After the run carefully remove the tube, and examine in the dark under UV light. At this stage two bands should be visible, an upper sharp band of chromosomal DNA and a lower diffused band of plasmid DNA.
15. Puncture the top of the tube and using a hypodermic syringe and needle, inserted through the side of the tube below the lower band, draw off the DNA of lower band and transfer to a sterile tube.
16. Remove the EtBr completely by partitioning aqueous phase several times with n - butanol saturated with TE.
17. Transfer the plasmid solution to dialysis tubing and dialyse the solution at 4ºC overnight against 2 litres of TE on a magnetic stirrer with 3-4 changes of the buffer.
18. Transfer the plasmid DNA solution to a suitable vial. Add 0.1 vol of 3 M sodium acetate solution and mix.
19. Add 3 volumes of ethanol mix well and keep at -20ºC for 2 hrs.
20. Centrifuge at 10,000 g for 10 min.
21. Decant the supernatant, wash the ppt with 70% ethanol.
22. Dry the DNA pellet in a vacuum desiccator.
23. Redissolve the preciptate in 200 µl of TE.

v) *In vitro* labeling of DNA by nick translation

In vitro labelling of DNA by nick translation is the main technique for producing radioactively labelled DNA. It is based on the ability of *E.coli* DNA polymerase-1 to add nucleotide residues to the 3'-hydroxyl terminus that is created when one strand of double stranded DNA molecule is nicked. In addition, the enzyme has a 5' to 3' exonuclease ctivity that can remove nucleotides from the 5' side and sequential addition on 3' side results in movement of the nick (nick-translation) along the DNA (Rigby *et al.*, 1977), radioactive DNA can later be used in hybridization to detect the presence of homologous sequences.

a) Nick translation

For the use of genomic library screening, Du Pont Kit used

(α -32P) dCTP	5 μl
Buffer	4 μl
dNTP mixture	4 μl
DNA insert	4 μl
H_2O	9 μl
DNA poly I 0.6 units / μl	2 μl
DNAase I (0.2 units / ml)	2 μl
Total volume	30 μl

Protocol

1. Set up nick translation reaction as follows in the Eppendorf tube (the volume vary with different nick translation kits)

i)	Quantity of DNA to be labelled (1 μg)	10 μl
ii)	Volume of nucleotide/buffer (sol. I)	10 μl
iii)	Volume of water	20 μl
iv)	Volume of (α-^{32}p) dCTP solution (50 II)	5 μl
v)	Volume of enzyme solution (sol. 2)	5 μl
	Total volume	50 μl

2. Incubate in a water bath at 15°C for 90 min
3. Terminate the reaction by adding SDS to the final concentration of 0.5% (i.e. add 2.5 μl of 10% SDS).
4. Heat at SP- 60°C for 5 min. Typically, optimum incorporation (20-40% incorporation of free label) is achieved.

b) Separation of labeled DNA

Separate the nick translated recombinant plasmid DNA from unincorporated labeled dNTPs by chromatography through a small column of Sephadex G-50, to be used as probe.

1. Prepare a Sephadex G-50 column in a 5 ml pipette plugged with a sterile glass wool (Add 3.0 g of Sephadex G-50 in 250 ml of column buffer and let it stand overnight at room temperature and pour into the column).
2. Apply the DNA sample to the column. Wash the tube with 100 µl of column buffer and apply to the column.
3. Elute using the column buffer with a 0.5 ml/fraction into different fractions in Eppendorf tubes.
4. Follow the progress of incorporated and unincorporated a-^{32}P/dCTP down the column using a hand-held mini monitor. The leading peak should be collected into Eppendorf tubes as it elutes from the column.
5. Denature of pooled DNA in a boiling water bath for 5 min before using it for blot hybridization.

vi) Southern hybridization

Reagents

1. Prehybridization buffer

   5 x SSC

   0.05 M Sodium phosphate buffer

   2.5 x Denhardt's solution

   2.5 mM EDTA

   0.6% SDS

   5.0% Dextran sulphate in sterile distilled water.

2. Washing solutions

   2 x SSC - 0.1% SDS

   1 x SSC

   0.5 x SSC

   0.1 X SSC and 0.1% SDS

Protocol

1. Place the baked nylon membrane in hybridization bag.
2. Calculate the volume of the prehybridization buffer according to the size of the membrane. Buffer has to be added at the rate of 70 µl/cm^2 of the membrane.

3. Prewarm prehybridization buffer at 65⁰C.
4. Calculate the amount of salmon sperm DNA to be added. It is added at the rate of 100 μg/ml of prehybridisation solution.
5. Take salmon sperm DNA in an Eppendorf, denature for 10 min in boiling water bath and put in ice immediately.
6. Add the prehybridization solution along with salmon sperm DNA to the hybridization bag.
7. Squeeze out the air bubbles carefully and seal the bag.
8. Place the bag in water bath maintained at 65⁰C and leave overnight or for minimum eight hrs.
9. Prepare probe as described by nick translating the whole plasmid carrying genomic DNA insert.
10. Denature the probe DNA in boiling water for 10 min and put in ice immediately.
11. Take out the bag from water bath, cut one corner, and add the probe.
12. Carefully remove air bubbles and seal the bag.
13. Allow hybridisation to continue for 20-24 hrs at 65⁰C.
14. Take out the bag, cut one corner and remove the buffer. Discard buffer from the bag into the bottle meant for the purpose carefully.
15. Prewarm the washing solutions at 65⁰C. Maintain one shaker at 65⁰C.
16. Wash the membrane at 65°C for 30 min with each of the solutions. Monitor the count on the filter by mini counter.
17. Wrap the membrane with Saran wrap and expose in dark the X-ray films, inside a casette meant for the purpose.
18. Allow exposure to continue at -70⁰C for 1 to 7 days depending on count on the filter.
19. Remove the X-ray film in dark and develop the auto-radiogram as described in the following section.

vii) Autoradigraphy

Readymade developer and fixer are available. Remove the film after exposing and develop it in developer for 3-5 min in dark. Wash the film in stop bath for 1 min and fix the film in fixer for 5 min in dark.

Wash the film under running water for 30 min and dry it at room temperature. Do not reload a moist or cold cassette. Dry before reusing.

Typically ^{32}P samples are covered with clear plastic wrap (Saran Wrap is recommended). This protects the casette and intensifying screen from contamination with penetrating radioactivity. The plastic wrap may block a significant amount of the ^{35}S signal. If ^{35}S is used as the radioactive source, do not cover with the plastic wrap unless sample is sufficiently radioactive to penetrate plastic wrap.

B. RAPD- Random Amplified Polymorphic DNA Analysis

This technique utilizes a single random oligomer primer to amplify genomic DNA taken as template. The amplified fragment profile is generated by agarose gel electrophoresis. The difference in the fragment profile of different genotypes can be used as genetic markers for variety or genotypic identification, genome mapping and gene tagging.

Requirements

1.	Taq DNA polymerase	5U/µl
2.	Random Decamer primers	15 ng/µl
3.	Deoxy nucleotide triphosphates	10 mM
4.	Taq DNA polymerase buffer	10 x
5.	TAE Buffer	50 x

Protocol

1. Prepare the reaction mixture by adding the following in an Eppendorf tube

• Water	20 µl
• Buffer	2.5 µl
• dNTPs (200 µm)	0.25 µl each
• Primer	1 µl
• Enzyme	1 µ
• DNA (Template)	20-25 ng

2. Mix gently and carry out the polymerase chain reaction in the thermal cycler with the following specifications:

   Cycle 1 92°C – 3.5 min.

   37°C – 1 min

   72°C – 2 min.

   40 cycles : 92°C – 1 min 37°C – 1 min

72^{0}C – 2 min.

Last cycle: 72^{0}C – 7 min.

3. Prepare 1.5% agarose gel in 1 x TBE buffer
4. Add 2.5 μl loading dye to each tube.
5. Load the samples and carry out electrophoresis at 40V for four hr.
6. Stain gel with ethidium bromide (0.5 μg/ml).
7. View gel under UV light.
8. Score the parents for polymorphism.
9. Observe the segregation of polymorphic bands in the F_2 plants.
10. Group the F_2 plants and find the frequency in each group.
11. Perform X_2 test to confirm Mendelian segregation ratio of 3:1.

Suggested reading

Beckmann J.S. and Soller M. (1986). Restriction fragment length polymorphisms and genetic improvement of agriculture species. Euphytica 35: 111-124.

Koundal Madhu, Sharma D.R., Koundal K.R. and Mohapatra T. (2005). Comparative evaluation of RAPD and AFLP based genetic diversity in brinjal. Journal of Plant Biochemistry and Biotechnology 14: 1-5.

Koundal Vikas, Yadava D.K., Arif Ali, Koundal K.R. and Mohapatra T. (2008). Evaluation of micro-satellite markers for genome mapping of Indian mustard (*Brassica juncea*). Journal Plant Biochemistry & Biotechnology 17: 69-72.

Rigby P.W., Diekmann M., Rhodes C. and Berg P. (1977). Labelling deoxyribonucleic acid to high specific activity *in vitro* by nick translation with DNA polymerase. J. Mol. Biol. 113(1): 237-51.

Williams J.G., Kubelik A.R., Livak K.J., Rafalski J.A. and Tingey S.A. (1990). DNA polymorphisms amplified by arbitrary primers are useful as genetic markers. Nucleic Acids Res. 18: 6531-6535.

19

Development and Characterization of Transgenic Crop Plants

Example: Isolation and Characterization of Protease Inhibitor Gene and Promoter from Blackgram (*Vigna mungo* L.) and Genetic Transformation and Molecular Analysis of Transgenic Tobacco

A. Introduction

The insect pest menace is the major factor that destabilizes crop productivity. One of the practical means of increasing crop production is to minimize the pest associated crop losses. Insect pest control by toxic chemicals has brought about considerable protection to crop yields over the past few decades. However, extensive and indiscriminate use of chemical pesticides has resulted in environmental pollution, adverse effect on human health and other organisms, reduction in beneficial insect population and development of resistance in insect pests. Plants have a wide array of defense proteins and secondary metabolites including the protease inhibitors and lectins to protect themselves from herbives from herbivores. Protease Inhibitors (PIs), which are ubiquitous in nature, are some of the prime candidates with highly proven inhibitory activity against insect pests. They are able to suppress the activity of enzymes in insects, phytopathogenic microorganisms and nematodes. These inhibitors form complexes with proteases, thereby inhibiting their proteolytic activity, and also serve as storage proteins in plant storage organs.

With the advent of recombinant DNA technology and genetic transformation techniques, it is now possible to identify and insert genes from any organism into the crop plants to confer resistance to insect pests without any biological barriers. Development and use of transgenic crops that express an insecticidal protein can be used, as a component of IPM programme to minimize the crop losses in an eco-friendly manner. The multiple benefits of transgenic crops include flexibility in terms of crop management, decreased dependence on insecticide and high quality end products. Considerable progress has been made in developing transgenic plants with *Bt* toxin genes from *Bacillus thuringiensis* in different crops. However, due to wide spread use of one or a few genes, there are distinct possibilities of development of resistance to *Bt* genes. Hence, there is a need to identify alternative genes for deployment through transgenic crops to control insect pests.

The isolation of a gene begins with the cloning of fragmented DNA of the organism into appropriate vector. We have to obtain a very large number of such recombinants which together contain a complete collection of all (or nearly all) of the DNA sequences that represent the entire genome of an organism. Such a collection from which we withdraw the desired clone is called a gene library or gene bank. Any DNA cloning procedure has the following four essential parts: (i) A method for generating DNA fragments. (ii) A reaction which joins foreign DNA to the vector, (iii) A means of introducing the artificial recombinant into a host cell in which it can replicate., (iv) A method of selecting or screening for a clone of recipient cells that has acquired the recombinant. The gene library is screened by nucleotide probe or antibody probe to find the appropriate gene.

Now, in principle any gene can be isolated by constructing and screening the gene libraries or by PCR amplification if the gene sequence is available. Transgenic plants are now generally produced in a laboratory by adding one or more desirable genes to a plant's genome, and the techniques frequently called genetic transformation. The recent techniques of plant genetic engineering have allowed scientists to introduce highly specific modification in transgenic plants. Broadly speaking three gene transfer techniques have been developed which are used for developing transgenic crop plants. These are: (i) *Agrobacterium* mediated gene transfer (ii) gene transfer by particle bombardment system and (iii) direct gene transfer to protoplasts through electroporation. Selectable marker genes or visible reporter genes need to be co-transformed with gene of agronomic interest. The objective of expression of selectable marker gene is essentially to identify the transgenic events among the large population of transformed cells. Such transgenic developed by the scientist are tested and evaluated for the novel parameters following the molecular analysis and target trait expression Using plant genetic engineering crops are genetically modified by manipulating plants own genes /and by introducing genes from unrelated organisms for developing inherent resistance to biotic and abiotic stresses which are major cause of low crop productivity.

The genes of plant origin have added advantages, as they are correctly transcribed, translated and processed in the host plant after genetic transformation without any codon biasness as the primary product of a gene; the protease inhibitors have practical advantages over other native chemicals, which are products of complex biochemical pathways. The PIs could play endogenous roles in environmental responses and development Furthermore, it has been demonstrated that mechanical wounding, insect chewing and microbial infection enhanced the level of PIs significantly in local as well as remote tissues. The numerous independent studies have revealed that there was an enhanced expression of PIs in response to abiotic stress. By transferring single defense related gene from one plant to another, either with their own promoter or with constitutive promoters, genetically modified plants can be readily obtained. A number of transgenic plants have been developed with PIs, which confer resistance to different families of insects. They form complexes with proteases, thereby inhibiting their proteolytic activity. The PIs have been used to enhance resistance to insect pests in transgenic plants due to their small size, abundance, stability and high specificity for a particular class of digestive enzymes of insects.

In plant genetic engineering, the introduced target gene is often required to be expressed at a desired level and in specific tissues or organs of the modified plant. The efficient use of transgenic technology for crop improvement necessitates the availability of novel promoters that are constitutive, regulated and is native to the plant system. The most widely used, and best characterized, constitutive promoter is the 35S promoter from cauliflower mosaic virus (CaMV), member of the *Caulimoviridae* family. The CaMV 35S promoter drives high, mostly constitutive, levels of gene expression in a wide variety of plants. However, when more than one copy of the promoter is present in the genome or when plants are infected with CaMV, the activity of this promoter may be decreased. This sensitivity to homology-dependent gene silencing limits its utility in gene stacking / pyramiding. Most commercial genetically modified (GM) crops around the world contain a viral CaMV 35S promoter to drive the transgenes. It is only now becoming clear that this promoter is not as robust as laboratory and glasshouse studies have suggested and its function is influenced by as yet undefined physiological and perhaps environmental factors. In addition, the constitutive promoters drive the expression of transgene is believed to have expression all over the plant which may create an environmental impact besides wasting plant energy. So, it is important in the long-term, to identify other plant gene promoters that can be inducible and tissue specific in expression. It is also desirable to have such promoters available for the next generation of transgenic crops so that different traits can be stacked driven by independent promoters, so as to avoid transcriptional gene silencing induced by DNA sequence duplication. Looking into these aspects and suitability of protease inhibitor gene and promoter as competitive candidates, the present investigation was envisaged for the development and analysis of genetically engineered tobocco plants.

B. Methodolgy

1. Construction of genomic DNA library of black gram

i) Isolation and shearing of genomic DNA of Black gram

About 5 - 6 gm of etiolated seedlings grown in vermiculite in dark were ground in liquid nitrogen and high molecular weight DNA (~30 - 50 Kb) of Black gram cv. uttara was isolated by the CTAB method (Doyle and Doyle 1990) with some modifications. After precipitation of the DNA in ice cold isopropanol, long strings of DNA were spooled out with a wide-bore pipette to obtain high molecular weight DNA. Each step was handling gently to avoid shearing of DNA. Around 25µg genomic DNA of black gram was randomly sheared by passing it through a 200 µl small-bore pipette tip for about 100 times.

ii) Agarose gel electrophoresis of sheared DNA

Sample with dye containing 0.25 per cent (W/V) xylene cyanol and 0.25 per cent (W/V) bromophenol blue was added to around 1-2 µl of sheared DNA and DNA fragments were resolved on a 20 cm agarose gel. For agarose gel electrophoresis of DNA, 0.8% agarose gel was prepared by melting 0.8 g of agarose in 100 ml of 1X TAE. The gel was poured in a gel casting tray with comb properly placed. Ethidium bromide at concentration 0.5µg/ml was added before pouring the gel.After cooling to room temperature, the gel was given a pre-run at 40 V for 10-15 min in 1X TAE running buffer. The voltage applied across the gel was 40 V/cm for 8-10 hour. *Hind*III digested λ DNA was used as marker. Ethidium promide at concentration 0.5 µ/ml was added before pouring the gel.

iii) End-Repair of the insert DNA

End repair step generated blunt-ended, 5'-phosphorylated DNA. The end-repair reaction was scaled up according to the amount of DNA available.

1. First, all the fallowing reagents were thawed and mixed thoroughly before dispensing and combined all reagents by placing on ice.

Sterile water	32 µl
10X End-Repair Buffer	8 µl
2.5 mM dNTP Mix	8 µl
10 mM ATP	8 µl
Sheared insert DNA (20 µg)	20 µl
End-Repair Enzyme Mix	4 µl
Total reaction volume	80 µl

2. The reaction mix is incubated at room temperature for about 45 minutes.
3. Finally, gel loading buffer was added and incubated at 70°C for 10 minutes to inactivate the End-Repair Enzyme Mix.

iv) Size selection of the end-repaired DNA

The DNA to be used in the cloning process appeared as a long smear, so size fractionation of the end-repaired DNA is done by running on a 20-cm long, 1% LMP (Low Melting Point) agarose gel at 30-35 V overnight. This electrophoresis was performed in absence of ethidium bromide. The DNA which would be cloned should not be exposed to UV light under any circumstances.

1. LMP agarose gel (1%) was prepared in 1X TAE buffer. A wide comb was used to be able to load sufficient DNA into the gel.
2. The DNA size markers loaded into each of the outside lanes of the gel and 100 ng of Fosmid Control DNA was loaded into each of the inner adjacent lanes of the gel. Then the end-repaired insert DNA is loaded in the lanes between the Fosmid Control DNA lanes.
3. The samples were resolved by gel electrophoresis at room temperature for overnight at a constant voltage of 30-35 V and do not included any DNA stain in the gel or in the gel running buffer.
4. Following electrophoresis, the outer lanes of the gel containing the DNA size markers, the Fosmid Control DNA, and a small portion of the next lane that contains randomly sheared end repaired genomic DNA were cut off.
5. The cut-off sides of the gel were stained with ethidium bromide, and DNA was visualized with UV light. Further, marked the position of the desired size DNA in the gel using a razor blade.
6. The gel was reassembled and a gel slice was excised that is 2 to 4 mm below the position of the Fosmid Control DNA, So that the DNA recovered from the gel was ≥ 25 Kb.
7. The gel slice was then transferred to a sterile, screw-cap tube for extraction using the GELase method.

v) Recovery of the size-fractionated DNA

1. The tared tubes were weighed to determine the weight of the gel slice(s). It was assumed that 1 mg of solidified agarose will yield 1 µl of molten agarose upon melting.
2. The GELase 50X Buffer was warmed to 45°C. The LMP agarose melted by incubating the tube at 70°C for 10-15 minutes and quickly transferred the tube to 45°C.
3. The appropriate volume of warmed GELase 50X Buffer was added to make 1X final concentration. Then carefully added 1U (1 µl) of GELase enzyme preparation to the tube for each 100 µl of melted agarose. The

melted agarose solution was kept at 45°C and gently mixed. Incubated the solution at 45°C for at least 1 hour.

4. Then reaction was transferred to 70°C for 10 minutes to inactivate the GELase enzyme.
5. The 500 µl aliquots of the solution were transferred into sterile 1.5 micro-centrifuge tube(s).
6. The tube(s) were chilled in an ice bath for 5 minutes. Then tubes were centrifuged at maximum speed (>10,000 rpm) for 20 minutes to pellet any insoluble oligosaccharides. The upper 90-95 per cent of the supernatant, which contains the DNA, was carefully removed into a sterile 1.5-ml tube.
7. Precipitated the DNA a) Added 1/10 volume of 3 M sodium acetate (pH.7.0) and mixed gently. b) Then 2.5 volumes of ethanol, mixed by gentle inversion. c) Allowed for precipitation to proceed for 10 minutes at room temperature. d) The precipitated DNA was centrifuged at 10,000 rpm for 20 minutes. e) The supernatant was carefully removed from the pelleted DNA. f) The DNA pellet was washed twice with cold, 70% ethanol, repeating steps (d) and (e), taking care not to disrupt the DNA pellet. g) After the second 70% ethanol wash the pellet was air-dried for 5 minuts. Gently resuspended the DNA pellet in TE Buffer.
8. Finally, the DNA concentration was determined by running an aliquot of the DNA on an agarose gel using dilutions of known amounts of the Fosmid Control DNA as standard.

vi) Ligation reaction

The following reagents were combined in the order listed and mixed thoroughly after each addition.

Sterile water	2 µl
10 X Fast-Link Ligation Buffer	1 µl
10 mM ATP	1 µl
Copy Control pCC2FOS (0.5 µg/µl)	1 µl
Concentrated insert DNA (0.25 µg)	4 µl
Fast-Link DNA Ligase	1 µl
Total reaction volume	10 µl

Incubated at room temperature for 4 hours and then transferred the reaction to 70°C for 10 minutes to inactivate the Fast-Link DNA Ligase.

vii) Packaging the copy control fosmid clones

1. On the day of packaging reactions, inoculated 50 ml of LB broth +10 mM $MgSO_4$ + 0.2% Maltose with 0.5 ml of the EPI300-T1^R overnight culture and flask incubated at 37°C with shaking to an A_{600} of 0.8-1.0 (~2 hours). The cells are stored at 4°C until needed in next step.
2. One tube of the MaxPlax Lambda Packaging Extracts was thawed on ice for every ligation reaction performed in the previous step. For example, one tube of the MaxPlax Lambda Packaging Extracts was thawed if the standard 10-µl ligation reaction was done.
3. When the extract was thawed, immediately transferred 25 µl (one-half) of each to a second 1.5 ml microfuge tube and placed on ice. Store remaining 25 µl of the MaxPlax Packaging Extract in a -70°C freezer for use in future step.
4. The 10 µl of ligation reaction was added from previous step to each 25 µl of the thawed extracts being held on ice in the tube.
5. Mixed by pipetting the solutions several times. Avoid the introduction of air bubbles. Briefly centrifuged the tubes to get all liquid to the bottom.
6. Then incubated the packaging reactions at 30°C for 2 hours.
7. After completion of 2 hour packaging reaction, the remaining 25 µl of Max Plax Lambda Packaging Extract was added from previous step to each tube.
8. Further incubated the reactions for an additional 2 hours at 30°C.
9. At the end of the second incubation, Phage Dilution Buffer (PDB) was added to 1 ml final volume in each tube and mixed gently. Then added 25 µl of chloroform to each tube mixed gently and stored at 4°C.

viii) Tittering the packaged fosmid clones

The titre of the phage particles (packaged Copy Control Fosmid clones) was determined as this will aid in determining the number of plates and dilutions required to obtain a library.

1. Serial dilutions of the 1 ml of packaged phage particles were made from previous step into Phage Dilution Buffer (PDB) in sterile microfuge tubes.
   a) 1: 10^1 Dilute 10 µl of packaged phage into 90 µl of PDB.
   b) 1: 10^2 Dilute 100 µl of the 1:10^1 dilutions into 900 µl of PDB.
   c) Dilute 100 µl of the 1:10^2 dilutions into 900 µl of PDB.

2. Added 10 ml of each above dilution, and 10 ml of the undiluted phage, individually, to 100 ml of the prepared EPI300-T1^R host cells from previous step. Incubated each tube for 1 hour at 37^{0}C.
3. Spread the infected EPI300-T1^R cells on an LB plate + 12.5 ìg/ml chloramphenicol and incubated at 37^{0}C overnight to select for the Cop Control Fosmid clones.
4. Colonies were counted and the titer of the packaged phage particles was calculated from previous step using fallowing formula.:

$$\frac{(\text{Number of colonies}) \times (\text{Dilution factor})}{\text{Volume of phage plated } (\mu l)} \times 100\,\mu l/ml$$

ix) Plating and selecting the copy control fosmid library

1. Based on the titre of the phage particles determined in earlier step, diluted the phage particles from previous step with Phage Dilution Buffer to obtain the desired number of clones and clone density on the plate.
2. Mixed the diluted phage particles from previous step with EPI300-T1^R cells prepared in earlier step in the ratio of 100 µl of cells for every 10 µl of diluted phage particles.
3. Incubated the tubes at 37^{0}C for 1 hour: The infected bacterial cells spread on an LB plate + 12.5 µg /ml chloramphenicol and incubated at 37^{0}C overnight to select for the Copy Control Fosmid clone.

2. Preparation of protease inhibitor probe for library screening

i) Isolation of recombinant plasmid DNA

Mung bean (*Vigna radiata*) protease inhibitor clone available in our lab was streaked from glycerol stock on a Luria Agar plate containing ampicillin (0.1 mg/ml) for obtaining single white colonies. It was then inoculated in 10 ml Luria Broth containing ampicillin (1 mg/ml) and incubated overnight 37^{0}C with vigorous shaking (200 rpm) in an orbital shaker incubator.

Plasmid DNA was isolated using QIAGEN Plasmid Mini Kit. The bacterial culture was spun down in an oakridge tube at 13,000 rpm for 5 min at room temperature. The pelleted bacterial cells were resuspended in 250 µl of Buffer P1 by vortexing and then transferred to a microfuge tube. Buffer P2 (250 µl) was added and mixed thoroughly by inverting the tube gently 4-6 times. Next, 350 µl of Buffer N3 was added and mixed immediately by inverting the tube 4-6 times. The contents were centrifuged for 10 min at 13,000 rpm using a table-top microcentrifuge at 4^{0}C. The supernatant was applied to the QIAprep spin column and centrifuged at 13,000 rpm for 30-60s. The flow-through was discarded.

The QIAprep spin column was washed by adding 0.5 ml Buffer PB and centrifuged for another 30s. The flow-through was discarded. The column was washed by adding 0.75 ml Buffer PE and centrifuged briefly. The flow-through was discarded and the column was centrifuged for an additional 1 min to remove residual wash buffer. DNA was eluted by adding 50 μl buffer EB at the centre of each QIAprep spin column kept in a clean 1.5 ml microfuge tube. The isolated recombinant DNA is further used for isolating the desired gene of interest.

ii) Restriction of recombinant plasmid DNA

The above isolated recombinant plasmid (3μl) was checked on 0.8% agarose gel for quality and quantity. The recombinant plasmid (5 μg) containing 0.5 kb amplicon was restricted with *Not*I (10 U/μl) in the presence of high salt buffer (10X) to a final volume of 20 μl (made up with sterile milli Q water) and incubated at 37^{0}C for 90 min. The restricted sample was electrophoresed using a low melting point (LMP) agarose gel (0.8%) to further carry out elution of insert DNA (Mung bean PI gene).

iii) Elution of the mung bean DNA

The insert DNA was eluted from the above LMP agarose gel by following the protocol of QIAquick Gel Extraction Kit. The desired DNA band (lower MW) was cut off with the help of a sharp scalpel blade under UV light. The cut gel pieces were transferred to sterile microfuge tubes and weighed. The Buffer QG was added in the ratio of 3:1 (300 μl buffer: 100 mg gel piece) and incubated at 50^{0}C for approximately 10 min with intermittent overtaxing till the gel slices were completely dissolved. The isopropanol (one gel volume) was added to the sample, mixed and the sample was applied onto the QIAquick column. The QIA quick column was centrifuged for 1 min and the flow-through discarded. Buffer QG (0.5ml) was added to the QIAquick column and centrifuged for 1 min. Buffer PE (0.75 ml) was then added to the QIAquick column and centrifuged briefly. DNA was eluted by using Buffer EB (10 mM Tris-HCl, pH 8.5). The eluted DNA fragment was utilized for probe preparation which was subsequently used for hybridization.

iv) Preparation of probe

The eluted (purified) insert DNA containing Mung bean PI gene was used for radiolabelled probe preparation. The probe was prepared by using DecaLabel™ DNA labelling kit (MBI, Fermentas), as per the instruction manual. Labeling of DNA by random oligonucleotide-primed synthesis is based on the method of Feinberg and Vogelstein (1983) and is a good alternative to nick translation for producing uniformly radioactive DNA of high specific activity. The reaction mixture was prepared by adding 32 μl of deionised water, 10 μl of Decanucleotide 5X buffer and 2 μl (0.5 mg) DNA. The mixture was incubated in a boiling water bath for 5-10 min and cooled in ice immediately. The sample was spun at 5000 x g for 1 min. Then 3 μl of Mix C, 2 μl of [a-^{32}P]- dCTP (Bhabha Atomic Research Centre, Mumbai, India) and 1 μl of Klenow exo- was added. The contents were spun and incubated at 37°C for 10 min and 4 μl of dNTP was added. The mixture was spun and incubated at 37°C for 5 min and reaction was terminated by adding 1 μl 0.5 M EDTA (pH 8.0). The radiolabelled fragment was then used for hybridization.

3. Screening of genomic DNA library for the isolation of protease inhibitor

The screening of genomic library for the isolation of black gram PI gene and promoter was done using Mung bean PI as probe according to the methods of Davis *et al*, (1986) and Sambrook *et al*, (1989).

i) Primary screening

The grids were made on the petriplates and the colonies were streaked with the autoclaved toothpick on fresh chloramphenicol containing LA plates. The plates were incubated overnight at 37°C. Then overnight grown colonies were inoculated in 96 well ELISA plates containing LB freezing media with chloramphenicol antibiotic (master plate and replica plate) and incubated overnight at 37°C with shaking. Master plates were stored for alignment of positive clones and replica plates were centrifuged at 10,000rpm for 5 min and pelleted colonies were dissolved in 20 μl of autoclaved milliQ water. About 5 μl of dissolved bacterial colonies were spotted onto Hybond N+ nylon membranes (Amersham) using 96 well grids plate. After air drying spotted nylon membranes subjected to colony hybridization. Colony hybridization is a method for hybridizing a radiolabelled probe to bacterial colonies containing recombinant fosmids allowing the selection of colonies that have fosmid with the desired insect. Colony hybridization and further processing were done as per the protocol of Davis *et al*, (1986).

ii) Pre-hybridization and hybridization

After air drying spotted nylon membranes were transferred with their DNA side facing outside to a hybridization bottle. Twenty five ml of pre-hybridization buffer (0.5 M sodium phosphate buffer pH 7.2, 7% SDS, 10 mM EDTA) was then added to the hybridization bottle so that each square cm area of blot received 100 μl of buffer. To avoid nonspecific binding, DNA of Salmon sperm (1 mg/ml) was added to the buffer. Pre-hybridization was done in a hybridization oven at 65°C for 2 hour. Excess of pre-hybridization solution was removed from the bottle, so that now it contained only 50 μl buffer per square cm of blot.

Radiolabelled probe prepared from mung bean PI clone was added to the pre-hybridization solution and hybridization was allowed to proceed for 16-18 h at 65^0C with continuous rotation.

iii) Washing of filters and autoradiography

After hybridization, the bottle was removed from the hybridization oven and hybridization solution was drained out carefully into a disposable bag. To remove free nucleotides and non-specifically bound probe, the filter was washed thrice sequentially in 200 ml of washing solution as follows:

1st wash : 2X SSC + 0.1% SDS at room temperature for 10 min.

2ndwash : 1X SSC + 0.1% SDS at 65^0C for 15 min.

3rd wash : 0.5X SSC + 0.1 % SDS at 65^0C for 10 min.

All washings were done with continuous shaking in a shaker. Autoradiography was done to locate radioactive DNA on the membrane. The blot was placed on a support, wrapped in a saran wrap and kept in an exposure cassette. An unexposed X-ray film (Kodak Lab Chemicals, USA) was kept on the blot in the dark room and cassette was closed tightly. It was wrapped in a black cloth to prevent light to enter. The film was exposed for 3 days at -80^0C. The film was developed in Kodak developer for 4 min at room temperature, washed in water. The only dark spots which showed convincing hybridization signals on the autoradiogram were marked on the filter after alignment with a marker pen and the reference number of each colony was noted down.

iv) Secondary screening

For secondary screening, each of the selected positive clones were inoculated in LB media with chloramphenicol antibiotic and recombinant fosmid DNA was isolated from overnight culture. Isolated fosmid DNA repackaged using MaxPlax Lambda Packaging Extracts and transfected with the host bacteria, and the titer was determined. An appropriate amount of phage was used to obtain 200 to 300 colonies in one small Petri dish (100 mm). Then colonies were inoculated in 96 well ELISA plates containing LB freezing media with chloramphenicol and incubated overnight at 37^0C with shaking. Overnight grown culture plates were centrifuged at 10,000 rpm for 5 min and pelleted colonies were dissolved in 20 μl of autoclaved distilled water. About 5 μl of dissolved bacterial colonies were spotted onto Hybond-N+ nylon membranes (Amersham) using 96 well grids plate. After air drying spotted nylon membranes subjected to pre-hybridization and hybridization process in the same way as in primary screening. The only dark spots which showed convincing hybridization signals on the autoradiogram which are truly positive clones were marked.

v) Confirmation of recombinant clones

After selection of positive clones, recombinant clones were inoculated in 10 ml LB broth supplemented with chloramphenicol and incubated overnight at 37⁰C with shaking. Fosmid DNA was isolated by alkaline lysis method as described earlier and was restricted with a combination of *Eco* RI and *Hind* III. The restricted DNA samples were run on 1% agarose gel, stained with ethidium bromide and photographed. The gel was blotted on nylon membrane, hybridized with mung bean PI probe and then autoradiographed. The sequencing of the inserts was carried out at the Chromous Biotech, Pvt. Ltd, India.

vi) DNA sequence analysis

Nucleotide sequence analysis of the sequenced clones was carried out using BLAST sequence alignment programme of NCBI database for homology search.

4. Alternative PCR method for isolation and cloning of protease inhibitor gene from black gram DNA

i) Designing of primers

The protease inhibitor gene specific forward and reverse primers were designed based on information available in the NCBI database by using primer online software. The conserved region among these genes evaluated through Clustal homology search. The primers were analyzed using Fast PCR software for self-dimmers, cross-dimmers with alignment, melting temperature calculation and also for priming efficiency. Primer alignment specificity was checked using NCBI blast. All oligonucleotides were commercially synthesized as highly purified salt-free products (Hysel, India). The lyophilized primers were reconstituted in commercially available TE buffer to reach a final concentration mentioned in the synthesis reports. For each gene, annealing temperature used was 5⁰C less than the T_m mentioned in the synthesis report of the primers. To facilitate directional cloning of protease inhibitor gene, *Kpn*I and *Xba*I sites were included in the forward and the reverse primers respectively. The nucleotide sequences of different forward and reverse primers are given below:

PIF : 5' GCGGTACCTTCAAAACTGTCATGAATTCAGC 3'

PIR : 5' CTCTAGATTAGTCATCATCTTTATCCATGG 3'

ii) PCR mixture

Amplification reactions were standardized with regard to primer concentration, primer annealing temperature and amount of template DNA. The reaction was set up as follows:

Vigna mungo genomic DNA	:	1.0 µl (30 ng)
Forward primer	:	1.0 µl (20 pmoles)
Reverse primer	:	1.0 µl (20 pmoles)
dNTP mix (10 mM)	:	0.5 µl (200 µM of each dNTP)
Total rection volume	:	2.5 µl
Taq DNA polymerase (3U/µl)	:	0.5 µl (1.5U)
Cfu DNA polymerase (2.5U/µl)	:	0.5 µl (1.25U)
Sterile double distilled water	:	18.5 µl
Total reaction volume	:	25.0 µl

iii) PCR conditions

Step 1	Denaturation	94°C	3 min
Step 2	Denaturation	94°C	40 sec
Step 3	Primer annealing	55°C	45 sec
Step 4	Primer extension	72°C	2 min
Step 5	Repeat steps 2 to 4	34 times	
Step 6	Final extension	72°C	10 min
Step 7	Hold	4°C	forever

iv) Checking of PCR products

About 5 µl of PCR amplification product was mixed with 1 µl of 6X loading dye and electrophoresed in 1.2% agarose gel. Gel was stained with 0.1% ethidium bromide staining and GeneRuler 1 kb DNA Ladder, ready-to-use was used as size marker. The UV image of the gel was captured using Alpha Innotech (Fluorchem® FC2) gel documentation system.

v) Purification of PCR products

PCR product was purified using QIA-quick PCR purification kit (QIAGEN), following manufacturer's instructions. The PCR reaction mixture was transferred to 1.5 ml micro centrifuge tube and 3 volumes of binding buffer PN were added. The solution was loaded on to a column and left at room temperature for 2 min. The column was placed in a collection tube and spun at high speed for 1 min in a microcentrifuge. The flow through was discarded and the column was placed back into the same collection tube. The column was washed by adding 750 µl of PE buffer. The flow through was discarded and the column was centrifuged for

1 min at 13,000 rpm. The column was placed into a fresh 1.5 ml microcentrifuge tube. To elute DNA, 50 μl of elution buffer was added to the centre of the QIA quick membrane, left at room temperature for 10 min and then kept over a fresh Eppendorf tube and centrifuged for 1 min at the maximum speed to collect the eluent into a fresh tube.

vi) Cloning into pGEM-T

Easy vector For PCR amplification *Taq* and *Cfu* thermo stable DNA polymerase were used. The *Cfu* DNA polymerase has proof reading function whereas *Taq* polymerase adds a single deoxyadenosine at the 3' ends of amplified fragments in a template independent fashion. This feature has been used to construct a series of vectors called pGEM-T Easy (Promega, USA) which carry a single 3' T-overhangs. The PCR products are easily ligated into this vector. The single T-overhang also prevents recircularization of the vector thereby increasing the efficiency of ligation. Also, this vector allows for alpha-complementation, hence, permits blue-white screening of the transformants. Also the polyclonal sites are flanked by *Eco*RI sites, thus allowing easy release of the insert.

vii) Ligation

pGEM-T Easy (Promega) cloning kit was used and the protocol was followed according to the manufacturer instructions. The ligation reaction for pGEM-T Easy was set up as follows:

Column purified PCR fragment (PI)	:	3 μl (200 ng)
pGEM-T Easy vector	:	1 μl (100 ng)
2 X ligase buffer	:	5 μl
T_4 DNA ligase enzyme (3 weiss U/μl)	:	1 μl
Total rection volume	:	10 μl

The ligation mixture was incubated overnight at 4^0C.

viii) Preparation of competent cells

E.coli competent cells were prepared by modified protocol of Mandal and Higa (1970). The *E.coli* strain DH5á was used. A single colony of the bacterium was inoculated into 5 ml S.O.B. (2% Peptone, 0.5% Yeast extract, 10 mM NaCl, 2.5 mM KCl, 20 mM $MgCl_2$) with nalidixic acid (10 μg/ml) and incubated at 37^0C overnight at 200 rpm.One ml of this starter culture was used to inoculate 100 ml S.O.B. with nalidixic acid (10 μg/ml) in a 500 ml sterile flask and allowed to grow at 37^0C with constant shaking (200 rpm) till an OD_{600} of 0.6-0.7 was attained. The cells were chilled for 30 min prior to centrifugation (specify centrifuge model/make and rotor) at 4000 rpm at 4^0C for 10 min. The bacterial pellet was resuspended in 10 ml (1/10th vol.) of ice-cold solution I and incubated in ice for 30 min. After centrifugation at 2500 rpm for 2 min at 4°C the supernatant was discarded and the pellet was resuspended gently in 2 ml of solution II. The 100 µl aliquots were dispensed into microcentrifuge tubes and plunged into liquid nitrogen for instant freezing and immediately transferred to -70°C freezer for storage until further use.

ix) Transformation of competent cells

The Competent cells were taken from -70°C freezers and thawed on ice for 20 min. 10 µl of ligation mixture was added to tubes and incubated in ice for 30 min. A heat shock was given by placing the tubes in a water bath at 42°C, exactly for 90 sec. and the tube was snap cooled in ice. After 5-10 min, 900 µl of LB was added to the tubes and was incubated at 37°C for 1 h with gentle shaking (150 rpm). The cells were pelleted by giving one minute spin at 3000 rpm and 150 µl of fresh LB was added to the pellet and cells were dispersed by gentle tapping. The cells were placed on agar plates containing LA + ampicillin (100 µg/ml) and spread on top with 25 µl of IPTG (0.1 mg/ml) and 25 µl of X-gal (20 mg/ml). The plates were inverted and incubated at 37°C overnight.

x) Selection of transformants and isolation of plasmid DNA

The transformed cells alone grew on LA plate with ampicillin as blue or white colonies. The recombinants (white colonies) were identified based on lack of alpha complementation of *LacZá*. White colonies, presumed to contain vector with insert (recombinant vector) were picked and inoculated in 10 ml LB with ampicillin (100 µg / ml) in sterile capped vials. Qiagen mediprep plasmid isolation kit was used for isolation of plasmid from selected white colonies and the protocol was followed according to the manufacturer instructions.

xi) Restriction analysis

Plasmid DNA was subjected to digestion with restriction enzyme, *Eco*RI to confirm the presence of insert in the vector.Restriction digestion was set up as follows:

Plasmid DNA	:	26 µl (1 µg)
*Eco*RI enzyme	:	1 µl (10U)
10 x Buffer	:	3 µl (1x)
Total rection volume	:	30 µl

A brief spin was given to settle the contents and the tube was incubated at 37°C for 3 hour.

xii) Agarose gel electrophoresis

5 µl of restricted plasmid was electrophoresed in 1% agarose gel along with 1 kb marker for confirmation of putative colonies. Gels were stained with 0.1% ethidium bromide and bands were visualized in a UV transilluminator. The UV image of the gel was captured using Alpha Innotech (Fluorchem[R] FC2) gel documentation system. The confirmed clones were sequenced at the Chromous Biotech, India.

5. Sub-cloning of protease inhibitor gene in a binary vector

PI gene fragment was excised from pGEM-T Easy clone with *Kpn*I and *Xba*I enzymes and inserted at the same sites in the binary vector pBinAR.

i) Double digestion of pGEM-T Easy and pBinAR vector

For release of PI gene insert-

pGEM-T Easy plasmid DNA	:	30.0 µl (3µg)
10 x Tango buffer	:	8 µl (2X)
*Kpn*I	:	1.0 µl (10U)
*Xba*I	:	1.0 µl (10U)
Total rection volume	:	40.0 µl

Double digestion of Vector-

pBinAR plasmid	:	30.0 µl (3µg)
10X Tango buffer	:	8 µl (2X)
*Kpn*I	:	1.0 µl (8U)
*Xba*I	:	1.0 µl (8U)
Total rection volume	:	40.0 µl

ii) Gel extraction

The insert cloned in pGEM-T Easy vector was released using restriction enzymes and eluted from gel for subcloning in pBinAR. The double digested pGEM-T clones were loaded in a 1.2% agarose gel and run at 70V for 1h. The agarose blocks with the desired inserts were sliced out using a sterile scalpel while viewing under the UV transilluminator. The gel slice was placed into a pre-weighed sterile microfuge tube. The protocol as given by the manufacturer (Qiagen, USA) was followed for eluting DNA from agarose gel, using Qiaquick Gel Extraction Kit.

Three volumes of buffer QG was added to the microcentrifuge tube containing gel slice and held in a water bath at 52°C for 10 min. The contents were vortexed briefly at short intervals to ensure complete melting of the agarose block. One volume of isopropanol was added and was put into the column supplied. The contents were centrifuged at 10,000 rpm for a min and the flow through was discarded. The column was washed with 500 µl of fresh buffer QG. A 750 µl of buffer PE was added and again centrifuged and the flow through was discarded. After placing a sterile microfuge tube underneath, the DNA was eluted out of the column by adding 40 µl of sterile double distilled water to the centre of the column. Double digested pBinAR was ethanol precipitated, dried and dissolved in 40 µl of sterile double distilled water.

iii) Ligation

Double digested pBinAR (*Kpn*I and *Xba*I)	: 5.5 µl (20 mg)
Double digested pGEM-T insert (PI)	: 15 µl (60 mg)
10 x T_4 DNA ligase buffer	: 2.5 µl
T_4 DNA ligase	: 1 µl (20 units)
Total rection volume	: 25 µl

The ligation was carried out overnight at 16°C in each case. The ligation product was used to directly transform competent *E. coli* cells and plated on LA plates containing required antibiotic (kanamycin 50 mg/ml) following the procedure mentioned earlier. The recombinant clones are used for plasmid for further confirmation.

iv) Plasmid isolation and restriction analysis

Plasmid was isolated using Qiagen Midi-prep plasmid isolation Kit. Double digestion of the isolated plasmid with appropriate restriction enzymes and 2X Tango buffer was done to confirm the putative colonies. The restricted plasmid was electrophoresed on 1% agarose gel along with 1 kb marker for confirmation of the size of insert.

6. Transformation of *Agrobacterium* using freeze thaw method

i) Preparation of *Agrobacterium* competent cells

A 100 ml of YEP was inoculated with 1 ml of an overnight culture of *Agrobacterium tumefaciens* strain GV3101 and incubated with vigorous shaking at 28°C. Culture was grown to log phase (OD_{550} 0.5-0.8) and cells were pelleted through centrifugation at 5000 rpm for 10 min at 4°C. The supernatant was discarded and cells were resuspended in 1 ml ice-cold 0.1 M $CaCl_2$, incubated on ice for 30 min and then dispensed into 100 µl aliquots in microfuge tubes, quickly frozen in liquid nitrogen and stored at -70°C.

ii) Transformation of *Agrobacterium*

The competent *Agrobacterium* cells were thawed on ice and 1 µg of plasmid DNA was added. The mixture was kept on ice for 5 min and then frozen into liquid N_2 for another 5 min, and then the mixture was incubated for 5 more min at 37°C in a water bath. 1 ml of YEP was added to each tube, sealed well and kept on a shaker (200 rpm) at 28°C for 4 hour. The cells were collected by spinning briefly in a micro-centrifuge and were spread on YEP agar plates containing kanamycin (50 µg/ml), rifampicin (25 µg/ml), and gentamycin (15 µg/ml) antibiotics. The plates were incubated for 2 days at 28°C. Liquid cultures (5 ml) were grown from the colonies and used for mini-preps. This plasmid DNA was used as the template for PCR with the original sequence specific primers for confirmation of putative colonies. The glycerol stocks were prepared from the cultures of confirmed colonies.

7. Development of transgenic tobacco

Agrobaterium cells containing the construct; pBinAR-PI was used for tobacco (var. *petit havana*) transformation using protocol mentioned by Hooykaas and Schilperoort (1992) with some modifications.

Agrobacterium tumefaciens strain GV 3101 transformed with construct pBinAR-PI was grown in YEMA (yeast extract mannitol agar) with, rifampicin (25 mg/ml) and kanamycin (50 mg/ml) at 28°C for 24 hours. Young leaves of tissue culture grown tobacco plants (2 to 3 week old plants) were cut into 1 cm square discs and infected with *Agrobacterium* culture treated with acetosyringone at 200 mM. Cultures were left for 20 minutes for agroinfection. These co-cultivated explants were blot dried on sterile filter paper and then placed on solid MS (Murashige and Skoog, 1962) medium without any hormones and incubated in dark for 2 days. The leaf squares were then transferred to solid MS medium containing 3% sucrose, 1 mg/l benzyladenine and 0.1 mg/l naphthalene acetic acid for callus induction and 200 mg/l cephotaxime to prevent further growth of *Agrobacterium*, kanamycin at 100 mg/ml was added to medium for selection. The cultures were incubated at 25°C under 16 hour photoperiod for 3 to 4 weeks. The shoots and excised calli were subcultured and incubated for another 4 weeks at the same culture conditions. Young shoots were then transferred to hormone free MS medium with 200 mg/ml cefotaxime and 200 mg/ml kanamycin simultaneously for selection of transformed shoots and for rooting. Cultures were incubated for 3 to 4 weeks and surviving plants were transferred to pots containing peat and acclimatized in glass house. The DNA was extracted from putative transgenic tobacco plants by CTAB method and used for molecular analysis.

8. Molecular analysis of T_0 and T_1 transgenic tobacco plants

Southern hybridization. Transgene expression analysis of PCR positive plants was done by a single step reverse transcriptase-polymerase chain reaction (RT-PCR) and through northern hybridization.

i) PCR confirmation of putative transgenic

The solation and purification of genomic DNA of putative transgenics plants was done as described earlier and used for PCR analysis. The fallowing sets of nptII and gene specific primers were used for PCR, primers for the nptII gene (Forward: 5' GATGGATTGCACGCAGG 3' and Reverse: 5' GAAGGCGATA GAAGGCG 3') and black gram PI gene specific primers (Forward: 5' GCGGTACCTTCAAAACTGTCATGAATTCAGC 3' and Reverse: 5' CTC TAG ATT AGT CAT CAT CTT TAT CCA TGG 3') were used to amplify 50 ng of genomic DNA from putative transgenics. The final volume of PCR reaction was 50 ul and amplification was done in 1X Taq buffer, 25 mM $MgCl_2$, 10 mM dNTP, 0.1 pM of each primer and 1 U of Taq DNA polymerase. Thermo cycling was performed in a thermo cycler (Biorad) as follows: an initial denaturation for 3 min at 94°C followed by 30 cycles of denaturation for 30 sec at 94°C, primer annealing at 55°C for both *npt*II and gene specific primers for 30 sec. & extension at 72°C for 30 sec. The PCR product was analyzed on 1% (w/v) agarose gel.

ii) Transgene expression analysis in transgenic plants

Transgene expression analysis of PCR positive T_0 and T_1 transgenic plants was done by semi-quantitative single step reverse transcriptase-polymerase chain reaction (RT-PCR).Total RNA from transgenics and control tobacco (non-transformed) plants was isolated by Trizol method. The 100 mg leaf tissue homogenized in 1 ml of Trizol reagent. Incubate the homogenized tissue for 5 min at room temperature to permit the complete dissociation of nucleo-protein complexes. Add 0.2 ml of chloroform, cap the sample tube and shake vigorously by hand for 15-20 sec and incubate at room temperature for 2-3 min. Samples were centrifuged at 12000 rpm for 15 min at 4°C. The mixture separates into lower red, phenol- chloroform phase, an interphase and a colourless upper aqueous phase. Aqueous phase containing RNA was transferred to a fresh tube and 0.5 ml isopropanol added to precipitate the RNA. Samples incubated at room temperature for 10 min and centrifuged at 12000 rpm for 10 min at 4°C. The RNA pellet washed once with 70% ethanol by centrifugation at 15000 rpm for

10 min at 4^{0}C, the pellet dried briefly in air and dissolved in RNase free water. The expression of PI gene was analyzed in transgenic plants through semi quantitative RT-PCR. The RNA so isolated was quantified using nanogrop at 260nm and integrity of RNA was evaluated by performing electrophoresis of denatured RNA on agarose gel (0.8%) containing formaldehyde. The RNA samples were diluted to 1 µg / µl concentration using DEPC treated water. An equal amount of total RNA from both transgenics and control tobacco plants was used for a one step RT-PCR reaction. This method avoids sample-to-sample cross contamination and since, all synthesised cDNA is utilized as a template for the PCR. 1ug of total RNA was used for RT-PCR by using PI gene specific primers (Forward: 5'-GCG GTA CCT TCA AAA CTG TCA TGA ATT CAG C -3' and Reverse: 5' -CTC TAG ATT AGT CAT CAT CTT TAT CCA TGG-3') & actin primers (Forward: 5' AGA CCT TCA ATG TGC CTG CCA TG 3' and Reverse: 5' GCT CAG CAG AGG TGG TGA ACA TG 3'). RT-PCR analysis was carried out using One Step RT-PCR kit (Qiagen) according to the manufacture instruction. Amplified product was size fractionated onto 1.0% agarose gel.

iii) Southern analysis of transgenics

Southern analysis of T_0 transgenics was done according to the method described by Sambrook *et al.*, 1989. Approximately 10 µg of genomic DNA was digested completely with *Bam*HI, electrophoresed in a 0.8% w/v agarose gel, and subsequently transferred to a Hybond N^+ nylon membrane (Amersham Biosciences, Buckinghamshire, UK). The 545 bp PI coding region was digested out from pGEMT Easy with *Eco*RI and used as a probe after labeling with [^{32}P] dCTP (BARC, Mumbai, India) using Deca Lebel™ DNA lebeling kit (MBI, Fermentas). Hybridization, incubation, washing, and detection of signals were performed as described earlier.

iv) Northern analysis of transgenics

Total RNA from T_1 transgenics and control tobacco (non-transformed) plants was isolated by Trizol method as described earlier. 20 µg of total RNA from each sample was run on 1% formaldehyde agarose gel and blotted onto Hybond N+ membrane (Amersham) according to the protocol of Davis (1986). Pre-hybridization was performed at 42^{0}C in the presence of 50% formamide. Black gram PI coding region was labelled with ^{32}P dCTP and used as probe. After hybridization, a membrane was washed sequentially at 42^{0}C by 2x SSC, 1x SSC/0.5% SDS & 0.5x SSC/0.1% SDS and autoradiagraphed as described earlier.

v) Segregation analysis of transgenics

Segregation analysis of T_1 seeds obtained from the self pollination of stable primary transformants (T_0) was done. Seeds were surface sterilized with 0.1% w/v $HgCl_2$ and allowed to germinate on half strength MS medium containing 50 mg/l kanamycin in order to determine the pattern of inheritance of the black gram PI gene. Non-germinated seeds were considered as sensitive. The surviving green plantlets were transferred to soil and grown in the glasshouse for further development. Segregation and inheritance of the black gram PI gene in randomly selected T_1 plants was tested through PCR using both *npt II* and black gram PI gene specific primers.

vi) Attached leaf bioassays

The efficacy of transgenics against insect attacks was tested using detached leaf bioassays with larvae of *Spodoptera litura*. The *S.litura* larvae were obtained from Division of Entomology IARI, New Delhi. Healthy leaves from 2 month-old wild type and T_1 transgenic plants were challenged with second instars of *S.litura*. The leaves were embedded in 3% agar-agar in 110 mm sterile petri plates following method of Sharma *et al.* (2005). This technique keeps the leaves in a turgid condition for about 1 week. The experiments were run in triplicates for each individual transgenic line and control plant. For each leaf placed in 110 mm sterile petri plates, 5 early second instars insects were released and initial weight of insects was recorded. The survival of insects within the petri plate was monitored at an interval of 24 h and insects were weighed after 3 days and 6 days of feeding. The bioassay was concluded upon maximum damage of the wild type (control) leaves. All larvae (both dead and surviving) were removed, their size recorded and also the extent of leaf damage was measured.

9. PCR based method for promoter isolation and sub-cloning in binary vector

i) Genome walker method for isolation of protease inhibitor promoter

Genome Walker DNA walking is a simple method for finding unknown genomic DNA sequences adjacent to a known sequence such as a cDNA and it was performed according to manufacturer instructions (Siebert *et al.*, 1995). Genomic DNA of black gram was isolated from etiolated seedlings and genome walker librarires was constructed (pools of uncloned adapter ligated genomic DNA fragments) by complete digestion of separate aliquots of DNA with *Dra*I, *EcoR*V, *Pvu*II and *Stu*I that leaves blunt ends. Thus developed library was used first for primary PCR using adopter primer1 and gene specific primer1, and then 1:50 diluted primary PCR product was used for secondary PCR using nested adopter primer 2 and nested gene specific primer 2. The desired band (PI promoter-II) from the gel was eluted and used for cloning into pGEM-T Easy vector as discussed earlier. To facilitate directional cloning of PI promoter-II *Hind*III and *Xba*I sites were included in the forward and the reverse primers respectively. The positive clones were sequenced and followed by annotation of promoter sequence for prediction of transcription start site and regulatory elements were done using GENESH online software, PLANT CARE and PLACE data bases.

ii) Sub cloning of protease inhibitor promoter in binary vector pBI121

The genomic clone containing 52 upstream sequence (PI promoter-I) was amplified by PCR and cloned into pGEM-T Easy vector as discussed earlier. To facilitate directional cloning of PI promoter-I *Hind*III and *Xba*I sites were added to the 5' end the forward and the reverse primers respectively. The nucleotide sequences of different forward and reverse primers are given below:

PIPF : 5' CCAAAGCTTTTCGATTCTAAACTGACGC 3'

PIPR : 5' GCATTGCATGTGAAAGGTGGCAGGCATG 3'

PI promoter-I (PIP-I) and PI promoter-II (PIP-II) fragments were excised from pGEM-T Easy clones with *Hind*III and *Xba*I enzymes and sub cloned into pBI121 by replacing *CaMV35S* promoter. The pBI121 binary vector was used as positive control.

iii) Double digestion of pGEM-T Easy and pBI121 vector

For release of PI promoters insert

pGEM-T Easy plasmid DNA	:	30 µl (3µg)
10X Tango buffer	:	8 µl (2X)
*Hind*III	:	1.0 µl (10U)
*Xba*I	:	1.0 µl (10U)
Total rection volume	:	40 µl

Removal of ^{35}S CaMV promoter of pBI121

pBI121 plasmid	:	30 µl (3µg)
10X Tango buffer	:	8 µl (2X)
*Hind*III	:	1.0 µl (10U)
*Xba*I	:	1.0 µl (10U)
Total rection volume	:	40 µl

The inserts cloned in pGEM-T Easy vector were released using restriction enzymes and eluted from gel by fallowing procedure mentioned earlier for subcloning into pBI121.Double digested pBI121 was ethanol precipitated, dried and then dissolved in 40 μl of sterile double distilled water.

iv) Ligation

Double digested pBI121 (*Hind*III and *Xba*I)	:	5.5 μl (20 ng)
Double digested pGEM-T inserts (PIP -I and II)	:	15 μl (60 ng)
10X T_4 DNA ligase buffer	:	2.5 μl
T_4 DNA ligase	:	1 μl (20 units)
Total reaction volume	:	25 μl

The ligation was carried out overnight at 16^{0}C in each case. The ligation product was used to transform competent *E. coli* cells and plated on LA plates containing required antibiotic (kanamycin 50 mg/ml) following the procedure mentioned earlier. The recombinant clones are used for plasmid for further confirmation.

v) Plasmid isolation and restriction analysis

Plasmid was isolated using QIAGEN Midi-prep plasmid isolation kit. Double digestion of the isolated plasmid with appropriate restriction enzymes was done to confirm the putative colonies. The restricted plasmid was electrophoresed on 1% agarose gel along with 1 kb marker for confirmation of the size of insert.

10. Transformation of *Agrobacterium* and development of transgenic tobacco plants

i) Transformation of *Agrobacterium*

The 100 ml of YEP was inoculated with 1 ml of an overnight culture of *Agrobacterium tumefaciens* strain LBA4404 and competent cells were prepared by fallowing previously discussed protocol.

The competent *Agrobacterium* cells were thawed on ice and 1 μg of plasmid DNA of each promoter was added and transformation was carried out by fallowing previously discussed protocol. Liquid cultures (5 ml) were grown from recombinant colonies and used for mini-preps. This plasmid DNA was used as the template for PCR with the original sequence-specific primers for confirmation of putative colonies and glycerol stocks were prepared from the cultures of confirmed colonies.

ii) Development of transgenic tobacco

Agrobaterium cells (LBA4404) containing the constructs; pBI121-PIP-I and pBI121-PIP-II were used for tobacco (*var.* petit havana) transformation using protocol, described by Hooykaas and Schilperoort (1992) with some modifications. The regenerated plants were hardened and then transferred to glass house of

National Phytotron Facility, IARI. New Delhi. DNA was extracted from putative transgenic tobacco plants by CTAB method and used for molecular analysis. Promoter expression study in transgenic tobacco plants was done by histochemical assays of GUS activity.

iii) Analysis of GUS reporter gene expression in transgenic tobacco

Histochemical analysis of *GUS* activity was done essentially as described by Jefferson *et al*, (1987). GUS reporter gene activity was visualized by histochemical staining with 5-bromo-4-chloro-3-indolyl-²-D-glucuronide (X-gluc, Sigma Chemical Company) as the substrate. All tissues including the leaves, root, stem and flowers were immersed as whole, while fruits were cut in transverse sections and incubated in GUS staining solution (0.5 mM X-gluc, 0.1 M NaHPO4 pH 8.0, 0.5 mM K3Fe(CN)6, 0.5 mM K4Fe(CN)6, 0.01 M EDTA pH8.0, 20% methanol and 0.1% Triton X-100) and incubated overnight at 37ºC. After staining, sections were rinsed in 75% ethanol for 2-3 times and photographed

C. Results

i) Construction and screening of genomic library for protease inhibitor gene

1. Isolation and size-fractionation of genomic DNA

For the genomic library construction high molecular weight DNA isolation of ~25-40 Kb is very crucial since that is the insert accommodation range of the Fosmid vector. For this purpose isolation of string-like DNA was ensured. After passing genomic DNA through a 200 µl small-bore pipette tip for 100 times the correct size sheared DNA of more than 25 Kb was obtained. For checking this, around 1-2 µl of sheared DNA was examined on a 20-cm agarose gel using the Fosmid Control DNA as a size marker (Fig. 1A).

The End repair of the insert DNA generated blunt-ended 5'-phosphorylated DNA. The size-fractionation of the end-repaired DNA was done by running on a 20 cm long, 1 per cent LMP (Low Melting Point) agarose gel at 30-35 V overnight. The electrophoresis was performed in absence of ethidium bromide. The desired size DNA (~25 Kb) was excised from the gel and extracted by GELase method.

2. Ligation of the DNA to fosmid vector and packaging the recombinant DNA

The gel extracted insert DNA (~25 Kb) was ligated with CopyControl™ pCC2FOS Fosmid vector. Genomic library construction depends on efficient ligation. Since successful ligation can only be determined after infecting the phages with the host bacteria, higher the number of cfu/ml of the library, higher is the efficiency of ligation. Using EPICENTRE® Biotechnologies Fast-Link DNA Ligase enzyme and buffer resulted in a titre of 3 x 10^5 to 1 x 10^6 cfu /ml, which was high. The recombined DNA was packaged using MaxPlax Lambda Packaging Extracts with high efficiency. Ligations were carried out at DNA concentrations of 0.2 μg / μl or more, which favours concatemers and efficient packaging. The care was taken ensuring that the colony forming units were the same for every dilution. The titre of colonies of black gram library was around 3 x 10^5 to 1 x 10^6 cfu /ml (Fig. 1B).

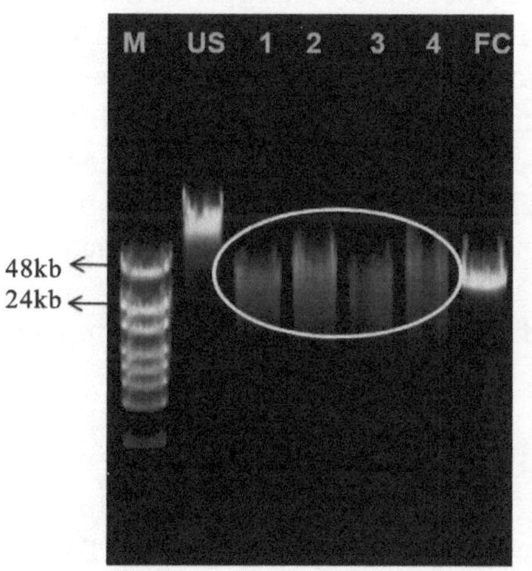

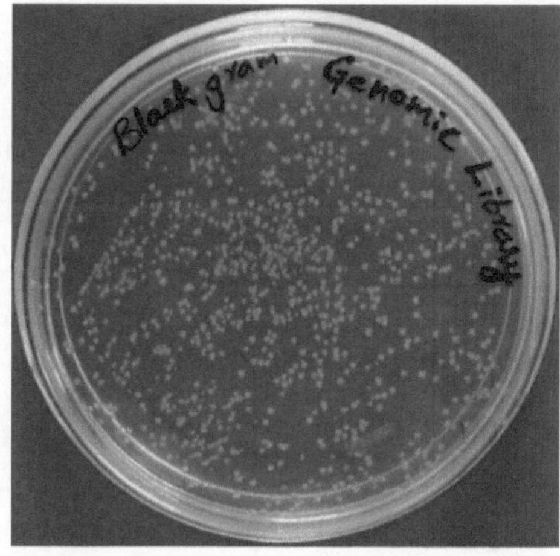

Fig. 1. Construction of genomic library of Black gram

A. Size fractionation of genomic DNA. M: High molecular weight DNA ladder, US: unshared genomic DNA, lanes: 1 to 4 sheared genomic DNA, FC: fosmid control DNA

B. Black gram genomic library in fosmid vector.

3. Screening of genomic library for protease inhibitor

The genomic library was screened by using mung bean PI heterologous gene specific probe. After primary screening positive signals were found and the reference number of each positive colony was traced out. The each positive colony lifted from the primary screening was passed through secondary screening with the same probe. After secondary screening each positive colonies were found to be very distinct and it was very easy to locate the individual positive colony. The positive clones were further confirmed by PCR using mung bean protease inhibitor heterologous gene specific primers and four PCR positive clones were obtained. The PCR positive clones sequenced at Chromous Biotech, Pvt. Ltd, India. Based on a haploid genome size of 5.74 x 10^8 bp, the coverage of the library is estimated to be 4.5 haploid genome equivalents, resulting in a

99.5% probability of recovering any specific sequence. The NCBI BLAST algorithm analysis showed that 1514 bp nucleotide sequence of genomic clone contains 5' upstream region and coding sequence of black gram protease inhibitor gene and it showed 93% homology with *Cicer arietinum* trypsin inhibitor. The 5' upstream sequence had transcription control regions like TATA signals between 624-631 bases, GAAT and CAAT signals between 498-554 bases. The longest ORF analyzed was between 746 and 1414 bases in the second reading frame and the ATG for this ORF starting at 746 bp. This ORF on translation gave a protein of 222 amino acids. The molecular weight of this protein was calculated to be 23.9 kD of which 21 amino acids were negatively charged, 16 positively charged with an isoelectric point of 5.0. The protein has an N- terminal signal sequence of 19 amino acids which is important in its translocation as has been reported in many proteinase inhibitors such as soybean, cowpea, pea and maize. This library will provide a good resource for isolation of many agronomical important genes and their promoters which can be further deployed for genetic transformation of field crops to various biotic stress resistances for enhancing the crop productivity. Protease inhibitors in combination with Bt-genes will be helpful for providing durable resistance to different biotic stresses in the transgenic crops resulting higher yield.

4. Alternative PCR based method of protease inhibitor isolation

A full-length open reading frame (ORF) of *VmPI* of black gram was obtained using a PCR-based approach and cloned into T/A cloning vector pGEM-T Easy. Its sequence consisted of 545 bp (NCBI GenBank accession No. HQ629949) (Fig. 2a). The ORF encoded a predicted polypeptide of 180 amino acid residues (Fig. 2b) with a calculated molecular mass of 20.2 kDa and an isoelectric point of 8.7. After ensuring sequence fidelity, it was compared with publicly available sequences by NCBI BLASTn search (http://www.ncbi.nlm.nih.gov) and analyzed by sequence alignment (BioEdit, USA). It showed 93% homology with *Cicer arietinum* trypsin inhibitor. The comparison with the other isolated protease inhibitors from different plants showed that the *VmPI* is basically a Bowman-Birk type proteinase inhibitor (BBI) of double headed (Fig.2c) type of plant serine protease inhibitors that block trypsin or chymotrypsin proteases. A ccomparison for conserved domain in the protein by NCBI- CD search showed that the regions from 113-168 amino acid residues have the conserved domain of which 9 residues compose reactive loops on the conserved domain of BBI similar to trypsin inhibitor Kunitz family of protease inhibitors. Southern blot analysis using the amplified *VmPI* gene fragment as a probe revealed hybridizing fragments of 3–4 different sizes, suggesting that several similar sequences are present in the black gram genome, indicating that *VmPI* belongs to an oligogene family in black gram.

>VmPI
TCAAAACTGTCATGAATTCACCCTAACTTGGCTTATAGTAATCGTGTAAGCA
AACAGACATGATATCGTGCATGCCTGCCACCTTTCACATGCAATGCAGAAA
GCAACAAAAAGATAACTATTCAGTGTTGCATCTGCAAATTTTATATAAACA
ACCCTCGTTGGGGCTCTTCTATATCTTCACTGCAAAAAGAAAATTTAGAAG
TTTGAACAAGAAGATGATGGTGCTAAAGGTGTGTGTGTTGGTAGTTTTCCT
TTTAGGGGTTACTGCTGCTGGCATGGATCTGAACCACCTGAGAAGTATTCA
TCATAATCATGACTCAAGCGATGAGCCTTCTGAGTCTTCAGAACCATGCTG
TGATTCATGCCGCTGCACTAAATCAATACCTCCTCAATGCCATTGTGCAGAT
ATTAGGTTGAATTCGTGCCACTCCGCTTGCAAATCCTGTATGTGTACACGAT
CAAGGCCAGGCAAGTGTCGTTGCCTTGACACTGATGATTTCTGTTACAAAC
CTTGCAAATCCATGGATAAAGATGATGACTAA

(a)

KTVMNSP-LGL—SCKQTDMISCMPATFHMQCRKQQKDNYSVLHLQILYKQ
PSLGLFYIFAKRKFRSLNKKMMVLKVCVLVVFLLGVTAAGMDLNHLRSIH
HNHDSSDEPSESSEPCCDSCRCTKSIPPQCHCADIRLNSCHSACKSCMCTRS
RPGKCRCLDTDDFCYKPCKSMDKDDD Stop

(b)

Fig. 2. Protease inhibitor gene isolated from Black gram
Nucleotide sequence (a), deduced amino acid sequence (b), 3D structure (double headed bowman birk type) (c), of protease inhibitor gene.

ii) Genetic transformation of tobacco plants

1. Development and molecular analysis of transgenic tobacco plants

In order to evaluate the functional significance of *VmPI* in transgenic tobacco plants, the plasmid designated as pGEM-T- *VmPI* was digested with *Kpn*I and *Xba*I to release 545 bp full lengths ORF and which was sub-cloned into the binary vector pBinAR under the transcriptional control of the constitutive cauliflower mosaic virus (CaMV) 35S promoter. The binary vector was mobilized into *Agrobacterium* and was used in transformation of tobacco employing the leaf disc co-cultivation method (Hooykaas and Schilperoort, 1992). A total of 10 putative T_0 transgenics were identified through genomic PCR based screening of which five independent transgenic plants were confirmed through different molecular approaches. The *VmPI* overexpressing plants contained one to several copies of the transgene(s) based on Southern blot analysis (Fig.4A) and single copy transgenic plants were identified for further analysis. The levels of *VmPI* transcript in T_0 transgenic plants were analyzed by PCR and reverse transcription (RT–PCR). The expression of a single *VmPI* transcript was observed in all the T_0 transgenic plants with varied expression levels (Fig.3A).

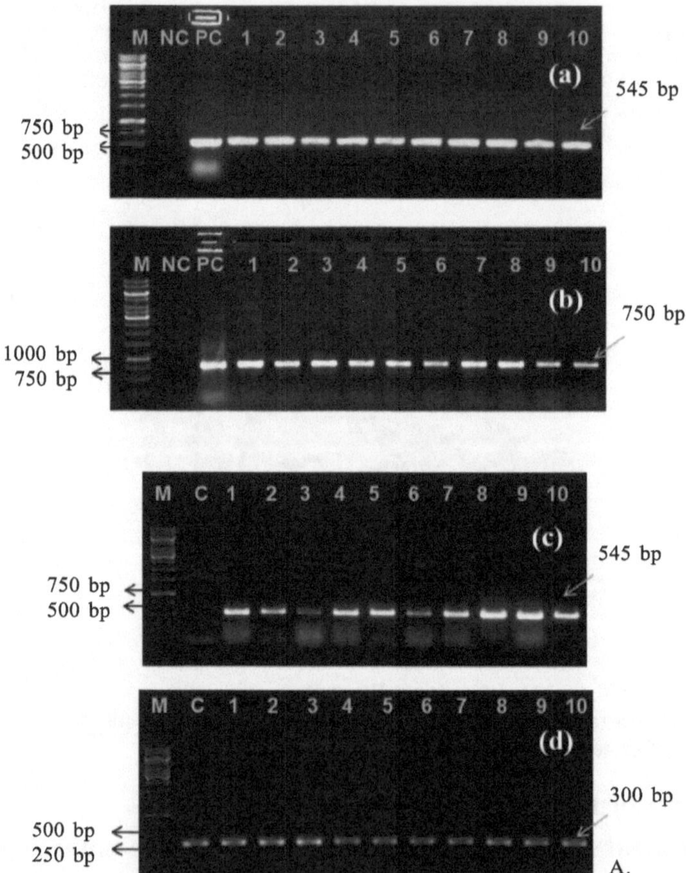

Fig. 3. PCR and RT-PCR confirmation of T_0 and T_1 transgenic tobacco plants.

A. PCR confirmation of T_0 transgenic plants, by gene specific primers (a), by *npt* II primers (b), M: 1kb ladder, NC: negative control, PC: positive control, lanes: 1 to 10 transgenic plants.

RT-PCR confirmation of transgenic plants by gene specific primers (c), by actin primers (d), M: 1 kb ladder, c: control, lanes: 1 to 10 transgenic plants.

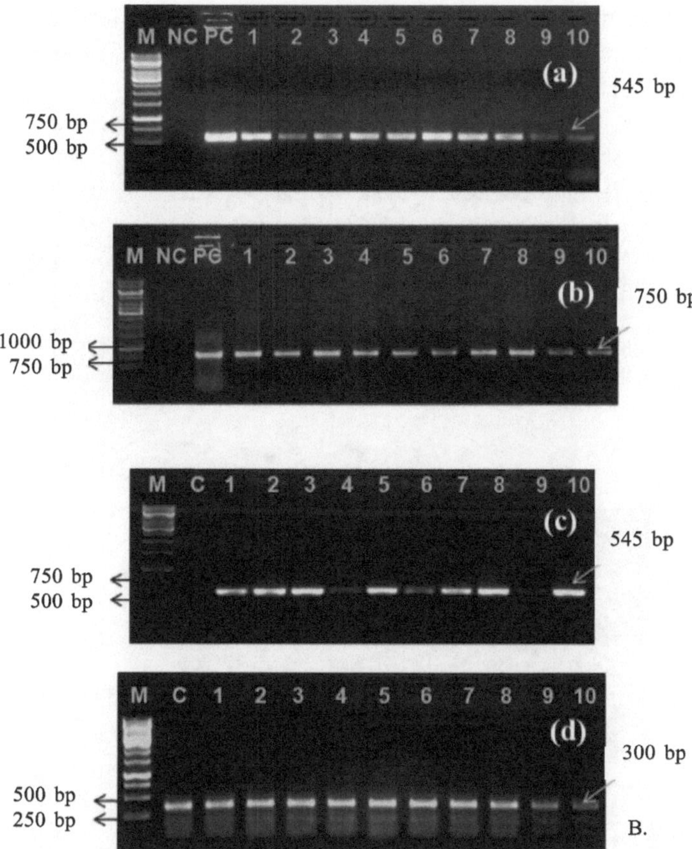

B. PCR confirmation of T_1 transgenic plants by gene specific primers (a), by *npt* II primers (b), M: 1 kb ladder, NC: Negative control, PC: positive control, lanes: 1 to 10 transgenic plants, RT-PCR confirmation transgenic plants by gene specific primers (c), by actin primers (d), M: 1 kb ladder, c: control, lanes: 1 to 10 transgenic plants M: ë *Hind* III marker, UR: unrestricted DNA, NC: negative control, lanes: 1 to 6 transgenic plants.

Transgenic plants 3 and 6 represented transgenics with a low expression level, while moderate expression was found in plants 2 and 7 and relatively higher expression was found in plants 1, 4, 5, 8, 9 and 10. The T_1 seeds were raised via self-pollination and T1 seeds were used in functional characterization.

Homozygous lines were identified on the selection medium containing 100 mg/l kanamycin in the T_1 generation. The T_1 transgenics were analyzed for the expression of *VmPI* in the selected lines by northern blotting (Fig. 4 B) and also by PCR and RT-PCR (Fig.3 B) and tested using detached leaf bioassay. The functional role of trypsin PIs is to inhibit the trypsin proteases that normally degrade proteins. The trypsin inhibitor activity was higher in T1 transgenic compare to wild type plants. There was significant difference in trypsin protease inhibitor activity (Fig.4c) between transgenic (line 5) and the wild type plants.

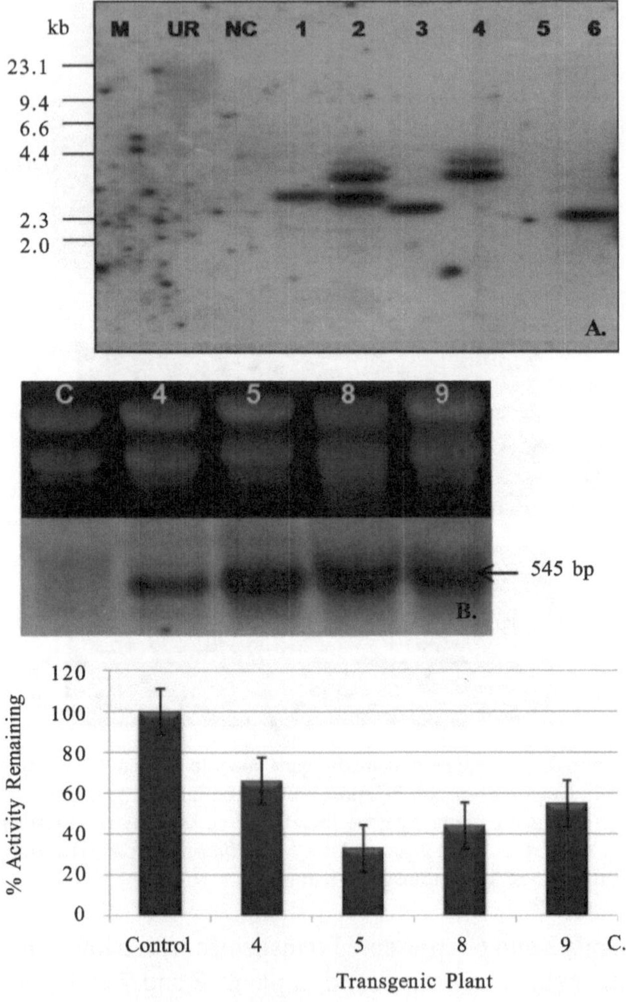

Fig. 4: Molecular analysis of transgenic tobacco plants.

A. Southern analysis of T_0 transgenic tobacco plants. M: ë *Hind* III marker, UR: unrestricted DNA, NC: negative control, lanes: 1 to 6 transgenic plants.

B. Northern analysis of T_1 transgenic tobacco plants. C: control plant, lanes: 4, 5, 8 & 9 transgenic plants.

C. Assay of protease inhibitor activities of T_1 transgenic tobacco plants.C: control plant, lanes: 4, 5, 8 & 9 transgenic plants.

2. Testing of insect resistance in transgenic

The insect resistance of T_1 plants expressing *VmPI* protein was tested using detached leaf bioassays with early second in star larvae of *S. litura*. Prominent differences in the amount of leaf area consumed by the larvae were noticed between the wild type and transgenic plants. The measurements of leaf area damaged and insect weight gained were taken at the end of the bioassay. The S. *litura* consumed the whole leaf in 6 days. High expression plants efficiently controlled second in star larvae of S. *litura*, while some damage was observed in the low expression plants. The leaf area damaged was negligible (Fig.5A,B) The larval growth was also arrested, and the weight of larvae grown on plants 5 and 8 was less than half that of the larvae feeding on wild type plants (Fig.5 C). Survival of larvae was significantly low in the high expression plants. High expression plants exhibited an elevated level of tolerance, and the larvae were stunted and did not appear healthy but the low expression plants behaved more like the wild type (Fig.5D).Therefore, the *VmPI* transgenics, which constitutively express the serine type trypsin PI, expected to be resistant to these pests, and the transgenic plants showed resistance to the second larval instars of *S. litura*. Under the elevated stress conditions in plants, there is degradation of proteins by proteases which might be repressed in transgenics due to constitutive expression of *VmPI.* The elementary function of a PI is to inhibit the function of proteases leading to stability of the respective proteins, which would otherwise be degraded by the proteases under biotic stress.

Control plant

Transgenic plant

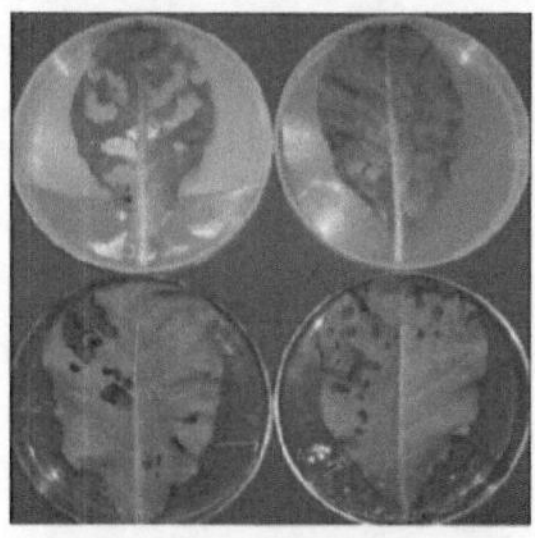

(a) 3 day after feeding

Control plant

Transgenic plant

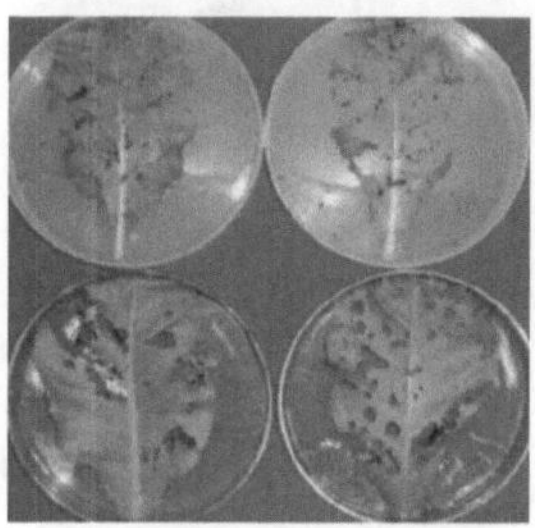

A.

(a) 6 day after feeding

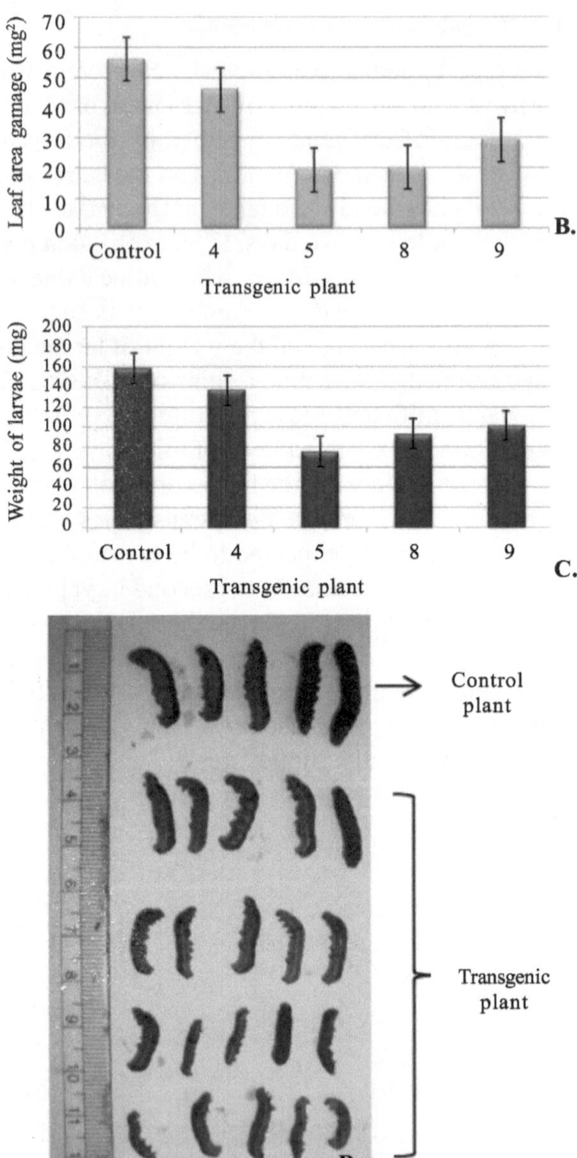

Fig. 5. Insecticidal bio-assay of transgenic tobacco plants.

A. Detached leaf insect bioassays using second instars larvae of *Spodoptera litura.*

B. The leaf area damaged after 6 days of feeding. C: control plant, lanes: 4, 5, 8 & 9 transgenic plants.

C. Weight of *S. litura* larvae after 6 days of feeding. C: control plant, lanes: 4, 5, 8 & 9 transgenic plants.

D. *S. litura* larvae after 6 days of feeding on control and transgenic plants.

iii) Isolation and expression of protease inhibitor promoter in tobacco

1. Isolation & analysis of the 5' regulatory regions of black gram PI gene

From genomic clone of black gram genomic library a 0.75 kb fragment containing the 5' upstream sequence of protease inhibitor gene (PIP) was amplified by PCR using a set of forward and reverse primers (Fig. 6A) and cloned into pGEM-T Easy vector. A sequence analysis showed a high similarity (93%) to the corresponding sequence (accession number AJ276262) in GenBank (Fig.6B) and this promoter belongs to Bowman Birk type of protease inhibitor promoter. The nucleotide sequence of Promoter cloned in this study was deposited at NCBI (GenBank: HQ629950). The promoter sequence was analyzed for transcription start site (TSS) and potential cis-acting transcription factor binding sites. Neural Network Promoter Prediction identified a TSS at 62 bp 5' to ATG. PLACE and Plant CARE analysis revealed a potential TATA box at 99 bp 5' to ATG and 36 bp upstream to the potential TSS. Several CAAT boxes, GAAT boxes, TTAGC plant defence response, GT1CONSENSUS light responsive elements, CACTFTPPCA1 cis-regulatory elements for mesophyll specific expression, MYB1AT dehydration responsive elements, an pollen specific expression element, CANBNNAPA endosperm and embryo specific and expression elements and wound responsive elements were found in the PIP promoter. Some of the relevant cis-elements detected and their relative positions from the translational start site ATG are specific for each kind of promoter.

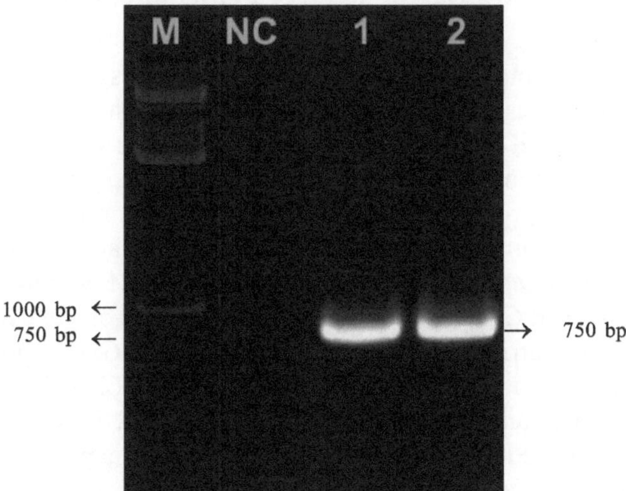

Fig. 6. Isolation of protease inhibitor promoter of Black gram.

A. Amplification of 5' upstream region of Black gram protease inhibitor gene from genomic clone. M:1 kb ladder, NC: negative control, lanes:1 & 2, 5' upstream region PCR amplicon from genomic clone.

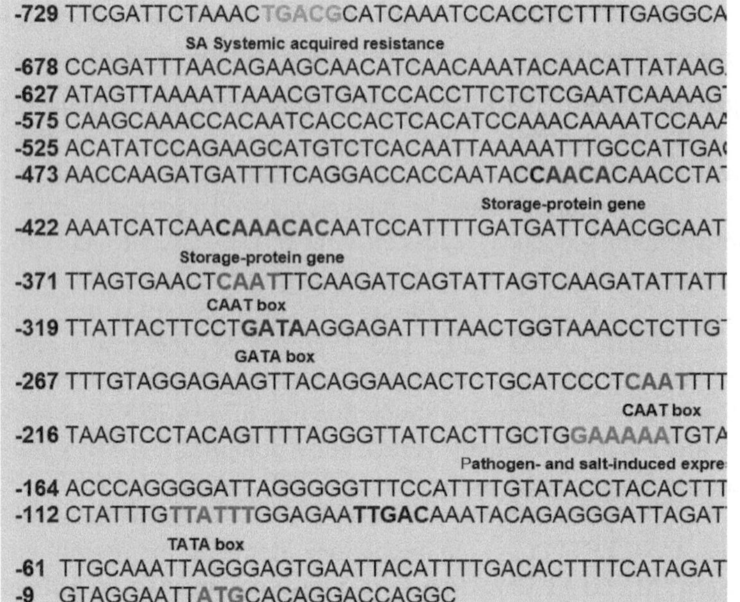

B. The 5' upstream sequence of Black gram protease inhibitor gene (750 bp) analyzed by PLACE and PLANT-CARE databases.

2. Development of transgenic tobacco plants and histochemical *GUS* assay

To characterize the putative PI Promoter in transgenic system, the 0.75 kb 5' flanking fragment was released from pGEM-T Easy clone by double digestion with *Hind*III and *Xba*I enzymes (Fig. 7A) and promoter::-glucuronidase (*GUS*) reporter gene fusion construct was prepared with same enzymes in binary vector pBI121. Then the promoter: β-glucuronidase (*GUS)* reporter gene fusion construct was transformed into *Agrobacterium* (Fig. 7B) and transgenic tobacco plants were generated (Fig. 8). Binary vector pBI121 having constitutive *CaMV35S* promoter: *GUS* reporter gene was used as control. Functional analysis of promoter was carried out by stable expression of *GUS* reporter gene in transgenic tobacco plant. In stable expression analysis, *GUS* activity was evident only in ovary, pistil and seeds of tobacco plant (Fig. 9). This qualitative *GUS* assay revealed that PI Promoter was able to drive the expression of reporter gene in different parts of tobacco plant. In conclusion, the promoter driven expression of gus showed the spatial and temporal regulation of the promoter in seeds of tobacco plants. The results of present study provide insights in to plant developmental processes and expression in seeds and other tissues in specific pattern indicate that the gene expression may have regulated according to cellular maturation processes. The isolation of tissue specific promoter and agronomically important genes (biotic and abiotic stress resistance) and their transfer to crop plants will be rewarding for the improvement of many field crops.It is expected that the development of transgenic crops using tissue specific promoters will be helpful for expressing genes in response to a particular stress resulting tolerance in plants to various stresses and increasing the crop productivity of field crops. As our sophistication in biotechnology improves, the need for more developmentally or environmentally regulated promoters has become evident and considerable effort is going on, to the discovery of specific tissue or biotic, hormonal, or abiotic stress responsive genes and promoters.

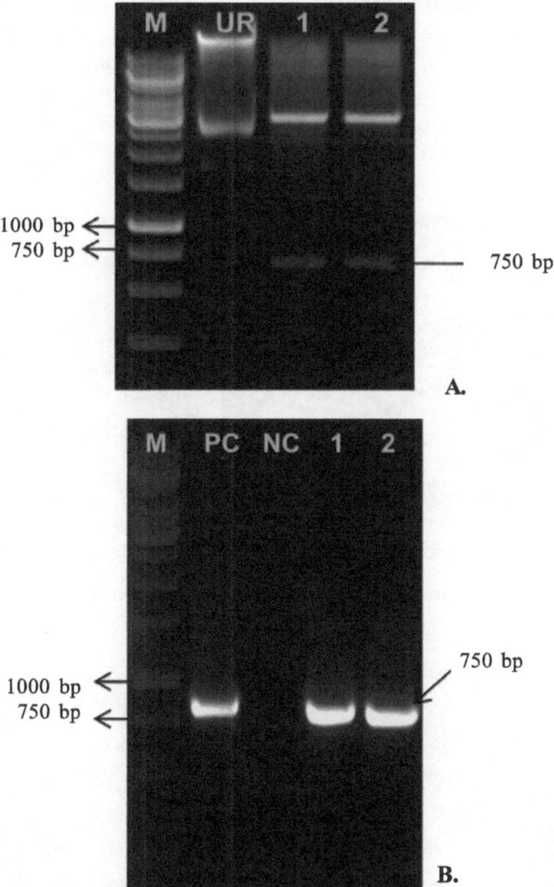

Fig. 7. Cloning of protease inhibitor promoter of Black gram.
A. pGEM-T-PIP clone digested with *Hind* III and *Xba* I M: 1 kb ladder, UR: unrestricted plasmid, lanes: 1 & 2: restricted plasmid clones.
B. PCR confirmation of *Agrobacterium* clones M: 1 kb ladder, PC: positive control, NC: negative control, lanes: 1 & 2: PCR amplicon of *Agrobacterium* clones.

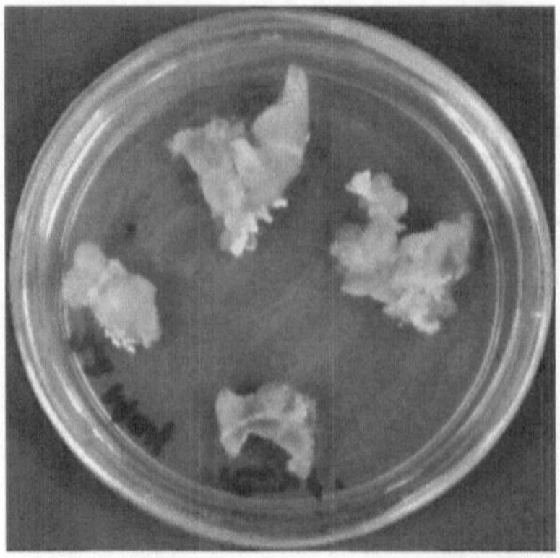

Agrobacterium Co-cultivation

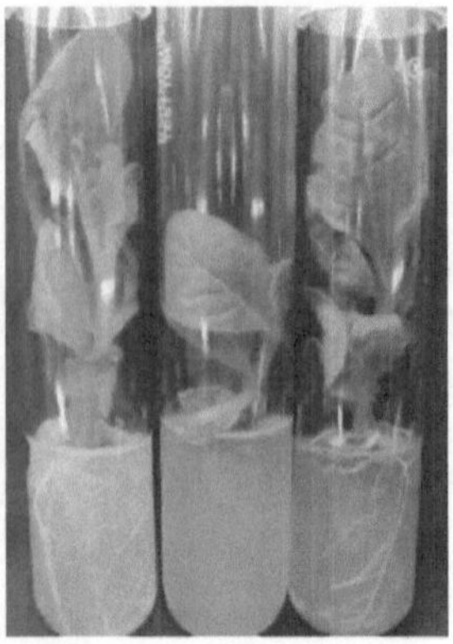

Shoot and root regeneration medium with Kan 200 mg/l

Plants growing under phytotron condition

Hardening of plants

Fig. 8. Different stages of development of transgenic tobacco transformed with Black gram protease inhibitor promoter

a) Promoter is expressing in ovary and pistil

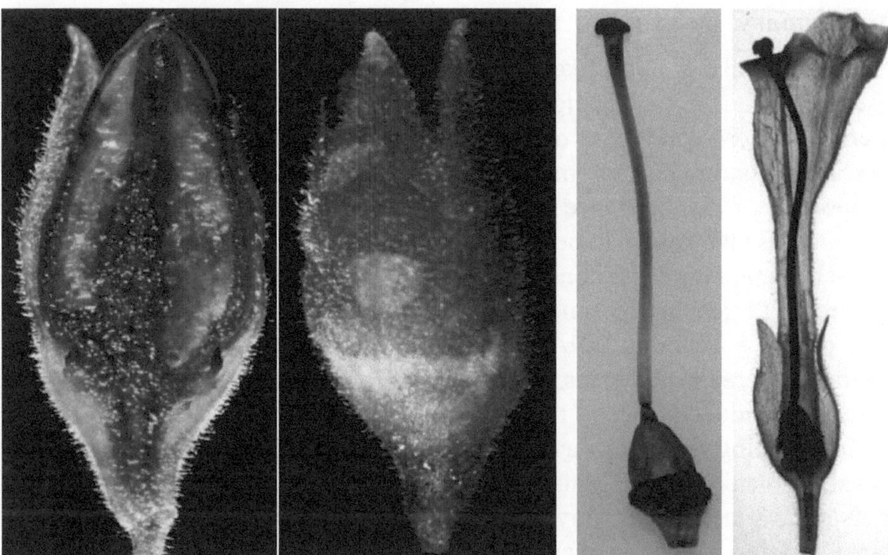

b) Promoter is expressing in ovary, pistil and seeds of tobacco

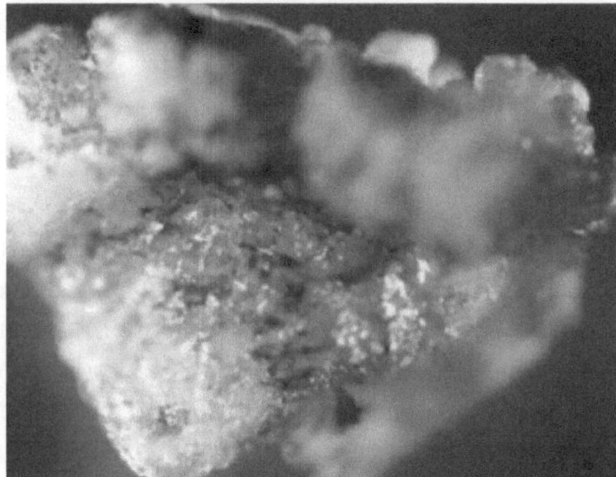

c) Positive control

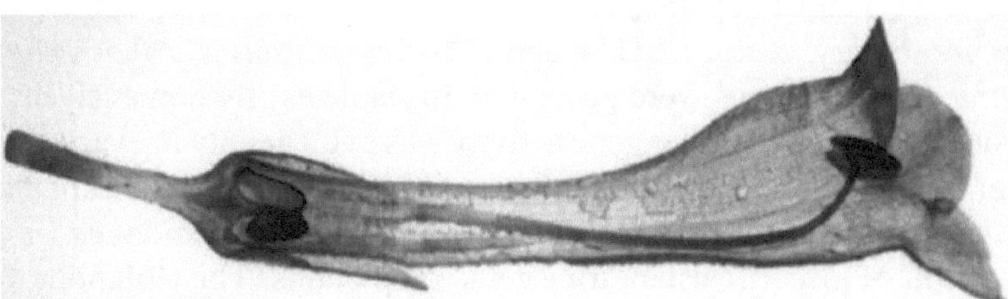

d) Negative control

Fig. 9. Functional analysis of promoter

D. Summary

Plant insecticidal proteins like protease inhibitors have been reported to confer insect resistance in transgenic plants and the use of PIs is regarded as a viable alternative to *Bt*- endotoxins or together both genes in insect pest control. Plant protease inhibitors (PIs) are small proteins which represent a class of well studied compounds of plant defence genes. Like other legumes, Black gram (*Vigna mungo* L.) is also known to be a source of plant PIs. The genomic DNA library of black gram *cv* uttra was constructed in Fosmid vector and recombinant clones containing PI gene and promoter sequence identified by using Mung bean PI gene probe. A genomic DNA library of Black gram (*Vigna mungo* L.) var. Uttara was screened by using Mung bean PI heterologous gene specific probe and a nucleotide sequence of 1415 bp having 5' upstream sequence and protease inhibitor coding sequence was obtained. The full length protease inhibitor gene was also isolated using PCR method. Using this protease inhibitor gene, tobacco transgenic plants were developed by *Agrobacterium* mediated genetic transformation.The molecular analysis of the transgenic plants at To and T1 stage was done by PCR, RT-PCR, Southern and northen hybridization. When confirmed transgenic tobacco (T1) plants were analyzed for insecticidal bioassays, the larval growth of *Spodoptera litura*, the tobacco cutworm was found severely retarded as compared to their growth on control plants. Therefore, Black gram protease inhibitor could be used as an alternative proteinase inhibitor can be used individually or in combination with other insecticidal genes for the development of insect-resistant transgenic crop plants.

In addition, a 0.75 kb 5' upstream sequence of protease inhibitor gene promoter (PIP) was amplified from genomic clone using a pair of primers and cloned into pGEM-T Easy vector. A sequence analysis showed a high similarity (93%) to the corresponding sequence in GenBank (AJ276262) and this promoter belongs to Bowman Birk type of protease inhibitors. The promoter sequence was analyzed for transcription start site and potential cis-acting transcription factor binding sites. Neural Network Promoter Prediction identified a TSS at 62 bp 5' to ATG and PLACE and PlantCARE analysis revealed a potential TATA box at 99 bp 5' to ATG and 36 bp upstream to the potential TSS. The PI promoter sub-cloned into binary vector pBI121 and transformed into *Agrobacterium*, and transgenic tobacco plants were generated. In stable expression analysis, the β-glucuronidase (GUS) reporter gene activity was evident only in ovary, pistil and seeds of tobacco plants. The black gram protease inhibitor and promoter could be used as individually or in combination with other insecticidal genes for the development of insect-resistant transgenic crop plants. The isolation of tissue specific promoter and agronomically important genes (biotic and abiotic stress resistance) and their transfer to crop plants will be rewarding for the improvement of many field crops. Therefore such strategy will be very useful for developing several transgenic field crops for developing durable resistance for various insect pests. A proposed management strategy for delaying insects′ development of resistance to plant protection transgenes, such as Bt toxins, is to deploy multiple insect control genes (such as PIs) with different modes of action in a single plant. There is evidence that the combination of PIs with a sublethal dose Bt toxin has a strong effect on the growth and development of insects. It is expected that the development of transgenic crops using tissue specific promoters and resistant genes will be helpful for expressing genes in response to a particular stress resulting tolerance in plants and thereby increasing the crop productivity of field crops.

Suggested reading

Davis L.G., Dibner M.D. and Batty J.F. (986). Basic methods in Molecular Biology, Elsevier, New York.

Doyle J.J. and Doyle J.L. (1990). Isolation of plant DNA from fresh tissue. Focus 12: 13-15.

Feinberg A.P. and Vogelstein B. (1983). A technique for radiolabelling DNA restriction endonuclease fragments to high specific activity. Anal. Biochem., 132: 6-13.

Hooykaas P.J.J. and Schilperoort R.A. (1992). *Agrobacterium* and plant genetic engineering. Plant Molecular Biology, 19: 15-38.

Jafferson R. (1987). Assaying chimeric genes in plants: the GUS gene fusion system. Plant Molecular Biology Reporter, 5(4): 387-405.

Mandal M. and Higa A. (1970). Calcium dependent bacteriophage DNA infection.

J. Mol. Biol., 53: 159-16.

Mudiyappanayar Jagdish and K.R. Koundal (2019). Construction and screening of Genomic DNA library of Black gram (*Vigna mungo* L.) for protease inhibitor genes. Indian J Agric. Biochem 32(1): 23-29.

Mudiyappanayar Jagdish and Koundal K.R. (2020) Constitutive expression of protease inhibitor gene isolated from Black gram (*vigna mungo L*) confer resistance to *sapodoptera litura* in transgenic tobacco. Indian Journal of Biotechnology 19(2): 94-101.

Murashige T., and Skoog F. (1962). A revised medium for rapid growth and bioassay with tobacco tissue cultures. Physiologia Plantarium, 15: 73–97.

Sambrook J., Fritsch E.F. and Maniatis T. (1989). Molecular cloning: A laboratory manual. Cold Spring Harbor Laboratory Press, New York.

Sharma H.C., Pampathy G, Dhillon M.K. and Smith, J.T.R. (2005). Detachedleaf assay to screen for host plant resistance to *Helicoverpa armigera*. J. Econ. Entomol, 98: 568 - 576.

Shasany Ajit Kumar and Koundal K.R. (2000). Isolation and characeterization of legumin promoter sequence from chickpea (*Cicer arietinum* L). Indian J. Expt. Biology 38: 36-46.

Siebert P.D., Chenchik A., Kellogg D.E., Lukyanov K.A. and Lukyanov S.A. (1995). An improved PCR method for walking in uncloned genomic DNA. Nucleic Acids Res. 23(6): 1087-88.

Colour Plates

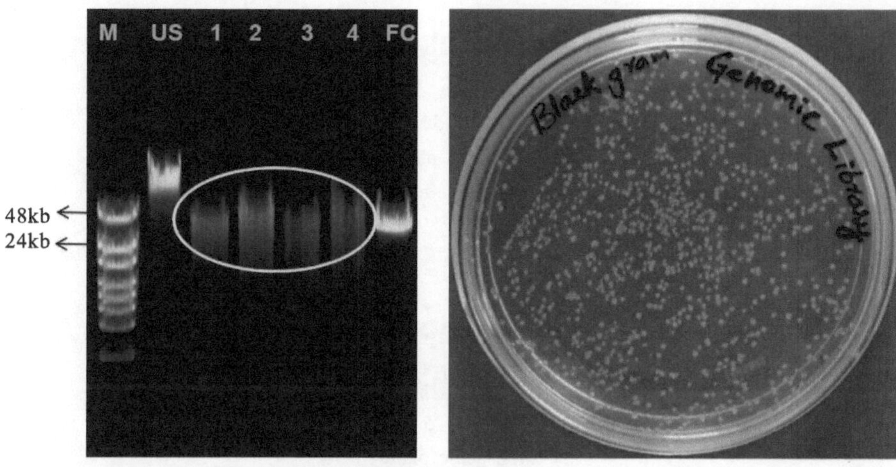

Fig. 1. Construction of genomic library of Black gram

A. Size fractionation of genomic DNA. M: High molecular weight DNA ladder, US: unshared genomic DNA, lanes: 1 to 4 sheared genomic DNA, FC: fosmid control DNA

B. Black gram genomic library in fosmid vector.

Fig. 2. Protease inhibitor gene isolated from Black gram

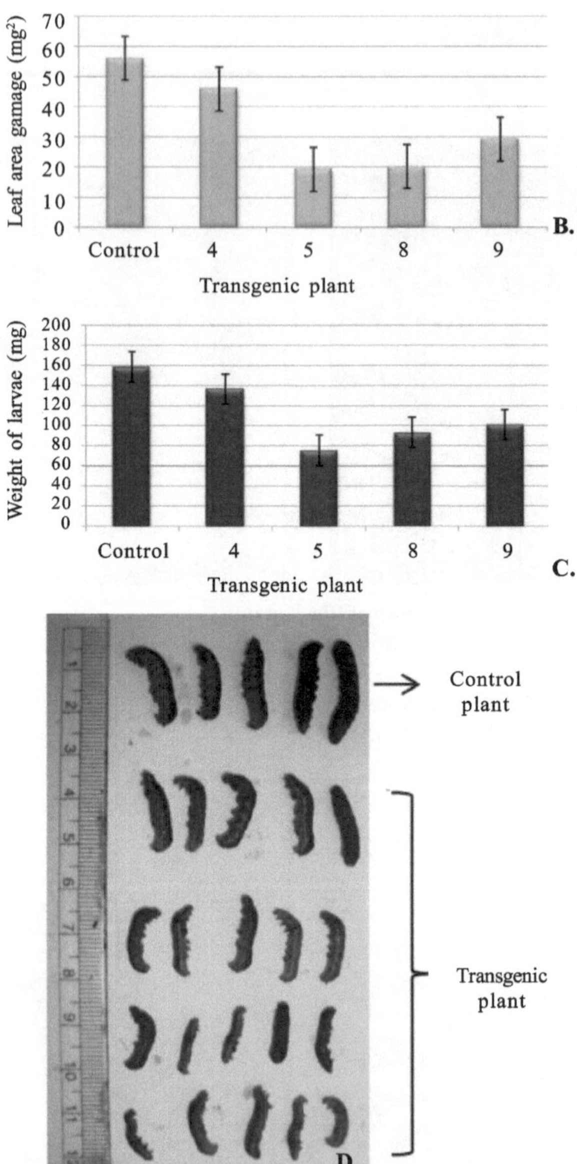

Fig. 5. Insecticidal bio-assay of transgenic tobacco plants.

A. Detached leaf insect bioassays using second instars larvae of *Spodoptera litura.*

B. The leaf area damaged after 6 days of feeding. C: control plant, lanes: 4, 5, 8 & 9 transgenic plants.

C. Weight of *S. litura* larvae after 6 days of feeding. C: control plant, lanes: 4, 5, 8 & 9 transgenic plants.

D. *S. litura* larvae after 6 days of feeding on control and transgenic plants.

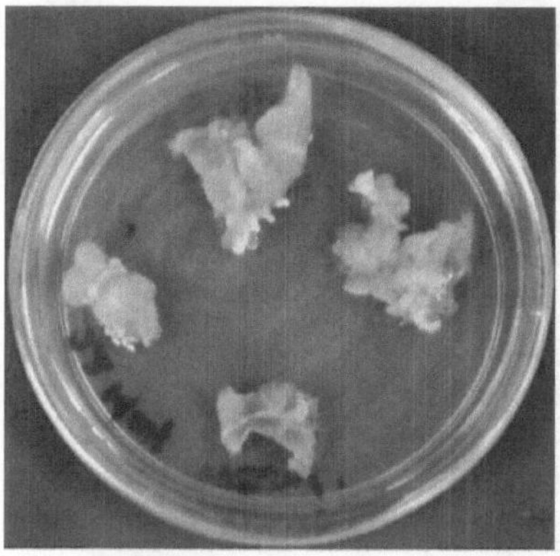

Agrobacterium Co-cultivation

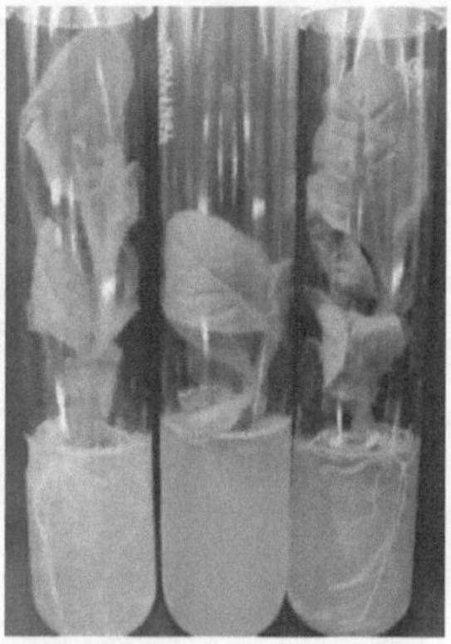

Shoot and root regeneration medium with Kan 200 mg/l

Plants growing under phytotron condition

Hardening of plants

Fig. 8. Different stages of development of transgenic tobacco transformed with Black gram protease inhibitor promoter

a) Promoter is expressing in ovary and pistil

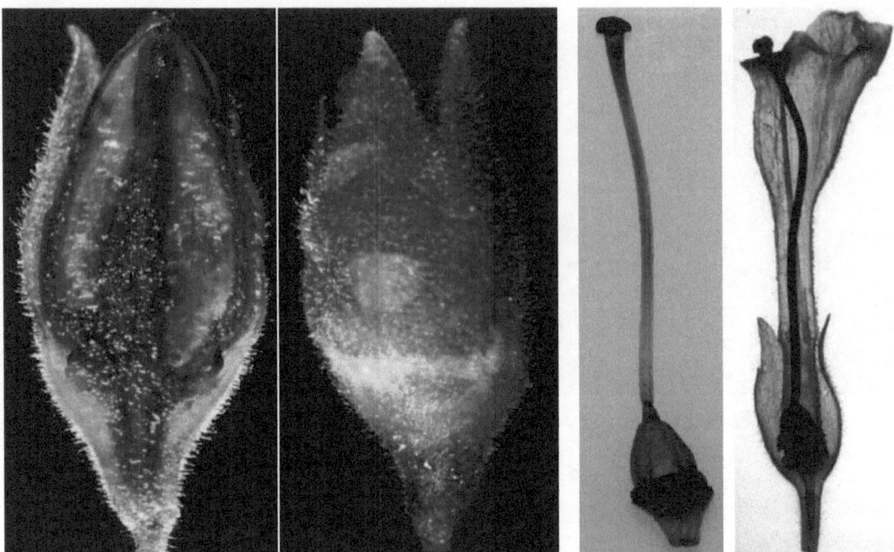

b) Promoter is expressing in ovary, pistil and seeds of tobacco

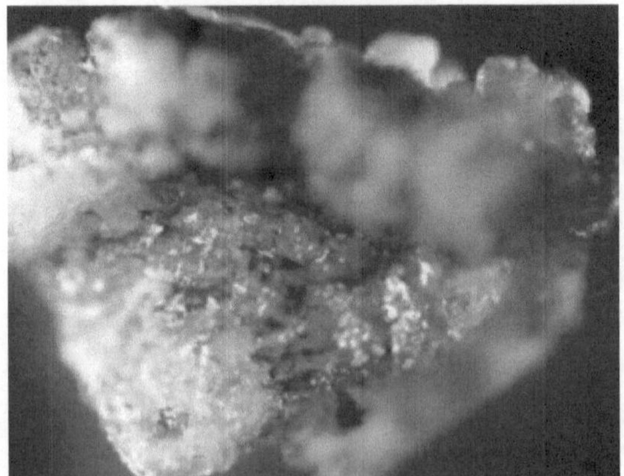

c) Positive control

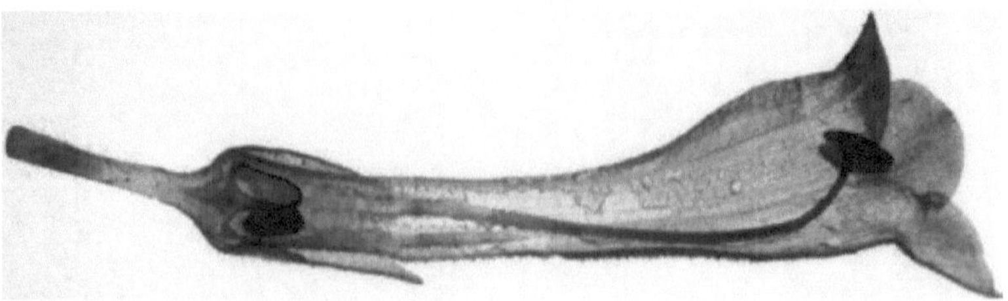

d) Negative control

Fig. 9. Functional analysis of promoter

Appendix

i) Storage of strains & plasmids

Storage of strains

Most cultures can be stored in stabs at room temperature. Unstable Tn5 insertion mutants, Hfrs, and strains carrying unstable genomes, are best stored at the low temperature (-70°C), Phage lysates are generally stable at 4°C over one drop $CHCl_{31}$ Both phage and bacteria can also be stored in 7% DMSO (or 15% glycerol) at 70°C. To remove phage or cells from storage, scratch the surface of the frozen material with a sterile needle. After streaking onto the appropriate medium or host strains, the vial can be refrozen at -70°C.

Storage of DNA and plasmids

Na_2EDTA chelates heacy metals which promote phosphodiester breakage. Ethidium bromide causes photooxidation of DNA with visible light and molecular oxygen and oxidation products of phenol can cause phosphodiester breakage.

The best temperature for storing DNA is 5°C. The DNA is exposed to frequent freeze-thaw cycles. causes extensive single and double strand breaks Storing at -70°C is excellent for long-term storage of DNA. It is best to store in high salt (>1M) and Na_2EDTA (>10 mM) at pH8.5 (Tris). Same way also plasmid DNA.

Storage of DNA is buoyant CsCl with ethidium bromide in the dark (wrapped in aluminium foil) at 5°C is excellent. There is about one phosphodiester break for 200 kb of DNA per year. Storage of DNA in the phage is better than storing the pure DNA. The DNA can be stored in the phage in buoyant CsCl for 5 years with no detectable breakdown (>one break/200 kb).

Ultraviolet light

UV light may be divided into three wave length groupings near UV (315-400nm), mid range UV (280-315nm) and far UV (200-280 nm). Maximal sensitivity in humans is at about 280 nm. Exposure to direct or indirect mid-range of for UV can cause acute eye irritation after a latent period 2-24 hrs. Because retina is sensitive to UV, eye damage may result without the subject being aware of the exposure. Skin is also sensitive to UV which may be a cause for skin cancer. Hence protect your eyes and skin from the effects of UV irradiation by wearing goggles with side shields by clothing, and by limiting exposure.

Preparation of phenol

All crystalline phenol must be redistilled at 160°C to remove contaamnants that cause or cross linking of DNA or RNA. Soon after distillation add 0.1% hydroxyquinoline. The melted phenol is extracted several times with oian equal volume of 1.0 M Tris pH 8.0 followed by 0.1 M Tris pH 8.0 and 0.2% β-mercaptoethanol, until pH of the aquenous phase in 7.6 phenol is stored in aliquots at 4°C under equilibration buffer for periods upto 1 month.

Phenol is widely used as a disinfectant and germicide. It is a dangerously toxic material that can produce poisoning when ingested, inhaled or absorbed through the skin. The toxic effect includes headache, nausea, and weakness, difficulty in breathing, unconcousness and death. Phenol is corrosive to skin, initially producing a softened area followed by servere burns. If phenol is spilled on other skin, flush immediately with large amount of water. Do not use ethanol. If eyes are conataminated, wash them with running water for about 15 min call for medical help.

^{32}P labeled compounds

β-particles with an energy of 1.71 MeV (6.1 meter range in air) is emitted by ^{32}P. Hence ^{32}P labeled compounds must be handled carefully with much caution using shields. When β particles hit targets, electromagnetic radiation known as Bremsstrahlung is produced, the yield of which is directly proportional to the density of material used for shielding. A second layer of high-density material may be added to absorb the Bremsstahlung emitted. Alwasys wear gloves (two pairs if necessary), protect eyes and use dosimeters. Always cover the work area with absorbant papers and use survey meter to check spillage. Eating, smoking or drinking while handling radioactive compounds should be banned. Use special tape to label containers and tubes in which radioactive materials are kept. The maximum permissible burden of ^{32}P is 30 μCi but the maximum permissible burden for bone is only 6μCi.

Selinization of plastic and glassware

Place the plastic and glassware to be silanized in desiccators. Add about 1 ml of dichlorodimethyl silane in a small container in a desiccators. Pull vacuum on the system until dichlorodimethyl silane boils. Close the system and allow sitting for about 2 hrs. Open the desiccators in a fume hood and allow to air oout several hours. Rinse the plastic and glassware with water and autoclave.

Common antibiotics used in the Lab:

	Stock solution		Working concentration (dilution)
	Concentration	Storage	
Ampicillin (sodum salt)	50 mg/ml in H_2O	-20°C	100 μg/ml (1:500)
Chlramphenicol	34 in mg/ml in ethanol	-20°C	170 μg/ml (1:200)
Kanamycin	10 mg/ml in H_2O	-20°C	50 μg/ml (1:200)
Streptomycin	10 mg/ml in H_2O	-20°C	50 μg/ml (1:200)
Tetracycline-Cl	5 mg/ml in ethanol	-20°C	50 μg/ml (1:100)

ii) Preparation of standard stock solutions

10 mM Ammonium Acetate

Dissolve 771 g of ammonium acetate (M.W. = 77.1 g/mole) on sufficient H_2O to make a final volume of 1 liter. Sterilize the solution by passing it through a 0.22 μm filter.

10 mg/ml BSA

Add 100 mg of BSA (Fraction V or molecular biology grade, DNase-free) to a 15 ml polypropylene tube containing 9.5 ml of H_2O.To reduce denaturation; always add protein to an aqueous solution instead of adding an aqueous solution to a protein. Gently rock the capped tube until the BSA is completely dissolved. Do not mix by vortexing (this causes foaming, which indicates protein denaturation). Adjust the final volume to 10 ml with H_2O. Divide into aliquots and store at -20°C. In general, this solution is not sterilized.

100x Denhardt's Reagent

Component and final concentration	Amount to add per 100 ml
2% Ficoll (Type 400)	2 g
2% Polyvinylpyrrolidone (PVP-40)	2 g
2% BSA (Fraction V)	2 g
H_2O to make 100 ml	

Dissolve the components in the H_2O. Filter to sterilize and remove particulate matter. Divide into aliquots and store at -20°C.

Divide into small aliquots and store -20°C. In general, this buffer is not sterilized.

1M DTT

The simplest approach for preparing a 1 M solution is to add 32.4 ml of H_2O to a 5 g bottle of DTT. Divide into aliquots and store at -20°C. This procedure avoids the weighing of this foul-smelling chemical.

Alternatively, transfer 100 mg of DTT (M.W. = 154.25 g/mole) into a microcentrifuge tube and add 0.65 ml of H_2O to make a 1 M solution. Sterilization is not required.

0.5 M EDTA (pH 8.0)	Dissolve 186.1 g of sodium salt of EDTA in 800 ml of double istilled water, adjust pH to 8.0 with 1N NaOH and make up the final volume to one liter with double distilled water and sterilize by autoclaving. Dissolve 233.8 g of NaCl in 800 ml of water and make up final volume to one liter with double distilled water and sterilize by autoclaving.
4 M NaCl	Dissolve 233.8 g of NaCl in 800 ml of water and make up final volume to one liter with double distilled water and sterilize by autoclaving.
IM Tris-Cl (pH8.0)	Dissolve 121.1 g of Tris-Cl salt in 800 ml of sterile double distilled water, adjust pH to 8.0 with 1N HCl and make up final volume to one liter with double distilled water and sterilize by autoclaving.
10% CTAB	Dissolve 10 g of CTAB powder in 80 ml of sterile double distilled water and make up the final volume to one 100 ml.
5 M Potassium Acetate	Add 11.5 ml of glacial acetic acid and 28, 5 ml 0f H_2O to 60 ml of 5M potassium acetate. The resulting solution is 3M with respect to potassium and 5M with respect to acetate.
10% SDS	Dissolve 100 g SDS in 900ml of H_2O, heat to 68°C to dissolve and adjust the pH to 7.2 with 1N HCL. Make up the final volume to 1 liter.
20XSSC	Dissolve 175.3 g of NaCl and 88.2 g of sodium bitrate in 800ml of H_2O and adjust the pH to 7.0 with 1N NaOH. Make up the final volume to 1litre and sterilise by autoclaving.
5X Loading buffer/dye	Gtycerol 50%, bromophenol blue, 0.0.25%, xylene cyanol 0.25%, EDTA 1 ml 1 M DTT. Dissolve 3.09 g of dithiothreitol in 200 ml of 0.01 M sodium acetate, pH 5.2, sterilize by filtration and dispense in 1ml aliquots, and store in -20C.
10xTBE (pH8.0)	Tris base 108 g, Boric acid 55 g, 0.5 M EDTA (pH 8.0) 20mIl, Dissolve in 1000 ml of deionizer water and autoclave.
10% Ethidium bromide	10 mg/ml in water. Wrap tube in aluminum foil and store at 4°C.
Gel loading buffer (6X)	Bromophenol blue 0.25%, Xylene cyanol FF, 0.25%, Sucrose 40% (w/v).
TAE buffer (50X)	Tris base 242 g, acetic acid 57.1 ml, 0.5 M EDTA (pH 8.0, 100 ml, Make up the volume to 1 liter. Autoclave.

iii) Preparatin of common laboratory solutions

DEPC – treated H_2O

Add 100 µl of fresh DEPC to 100 ml of H_2O to make a final concentration of 0.1% (v/v). Incubate at 37°C for at leasat 12 hours and then autoclave at 15 psi on liquid cycle for 20 minutes to inactivate the remaining DEPC.

70% (v/v) Ethanol

To prepare a solution of approximately 70% (v/v), mix 70 ml of absolute ethanol with 30 ml of sterile H_2O. Do not autoclave. Prepare as needed or store at -20°C. Sterilization is not required.

Formamide (Deionized)

Reagent-grade formamide can often be used directly, but it decomposes to formic acid and ammonia during storage. Deionized formamide (sometimes called molecular biology grade) can be purchased or prepared as needed. Do not use batches of formamide that have a yellow colour.

To deionize formamide, add Dowex XG8 mixed-bed resin to formamide in a glass breaker and stil gently with a magnetic stirring bar for 1 hour. Filter through Whatman No.1 paper to remove the resin. (Dowex XG8 resin can be purchased with a pH-sensitive blue indicator that becomes yellow upon removal of the acid from the formamide). Divide into small aliquots and store under nitrogen (to prevent oxidation) at -80°C. Sterilization is not required.

80% (v/v) Glycerol

Carefully measure 80 ml of glycerol in a non-wetting plastic graduated cylinder and transfer it into a 250 ml glass bottle or flask. Add 20 ml of H_2O and stir with a magnetic stirring bar until homogeneous. Sterilize by autoclaving.

Phosphate Buffers

Mixing 1 M NaH_2PO_4 (monobasic) and 1 M Na_2HPO_4 (dibasic) stock solutions in the defined volumes results in 1 liter of 1 M sodium phosphate buffer of the desired pH. To prepare the 1 M stock solutions, dissolve 138 g of $NaH_2PO_4.H_2O$ (monobasic; M.W. = 138 g/mole) in sufficient H_2O to make a final volume of 1 liter and dissolve 142 g of Na_2HPO_4 (dibasic, anhydrous; M.W. = 142 g/mole) in sufficient H_2O to make a final volume of 1 liter.

Chloroform: isoamyl alcohol (24:1)

A mixture of chloroform and isoamyl alcohol (24:1, v/v) is generally used to deproteinise nucleic acid preparations. Chloroform denatures the proteins and isoamyl alcohol helps to reduce foaming during extraction as well as helps to separate the queous phase.

5x TBE buffer

Component and final coincentration	Amount to add pere 1 liter
445 mM Tris base	54 g
445 mM Borate	27.5 of boric acid
10 mM EDTA	20 ml or 0.5 M (H 8.0)
H_2O	To make 1 liter

TBE can be prepared as a 5x or 10x stock buffer, but the 10x stock buffer will precipitate during storage. The 1x buffer (pH 8.3) is 89 mM Tris, 89 mM borate, and 2 mM EDTA.

1% Bromophenol Blue

Add 1 g of the H_2O soluble sodium form of bromophenol to 100 ml of H_2O. Still or mix by vortexing until fully dissolved. In general, this solution is not sterilized.

10% Ethidium Bromide (10 mg/ml)

Since this chemical is commonly used in a molecular biology laboratory, a 100 ml stock solution can be prepared. Carefully weigh 1 g of cthidium bromide, avoiding dispersal of the powder. Transfer into a wide-mouth bottle and add 100 ml of H_2O and a magnetic stirring bar. Still until dissolved. Wrap with foil and store at 4^{0}C.

Use of ethidium bromide powder can be completely avoided by purchasing commercially available solutions or by dissolving one 100 mg tablet of ethidium bromide (Sigma E 2515) in 10 ml of H_2O. Sterilization is not required.

1% Xylene Cyanole FF

Dissolve 1 g of xylene cyanole FF in sufficient H_2O to make a final volume of 100 ml. In general, this solution is not sterilized.

Gel Loading Solutions

Gel-loadikng solutions added to DNA samples for analysis on agarose or acrylamide gels can contain sucrose, glycerol, or Ficoll as the agent to increase the density of the sample. A dense sample drops evenly to the bottom of a gel well. Individual preference dictates which to use.

Tracking dyes indicate the extent of electrophoresis. Bromophenol blue migrates at the position of an approximately 300 bp linear double-stranded DNA and xylene cyanole FF migrates at the position of a 4 kb linear double stranded DNA in 0.5x TBE buffer. The preparations below use dye makers at a final concentration of 0.15 – 0.25%. In general, these gel-loading solutions are not sterilized. Add 1 μl of 10 x buffers to 9 μl of total sample.

iv) Common conversion of nucleic acids and proteins

1. Spectrophotometric conversions and purity of nucleic acids

   A_{260} unit of ds DNA = 50 ug/ml

   A_{260} unit of ss DNA = 33 ug/ml

   A_{260} unit of ss DNA = 40 ug/ml

   A_{260} unit of oligonucleotides = 20-30 ug/ml

   Pure DNA has an A_{260}/A_{280} ratio of 1.8 – 2.0 in Tris-Cl, pH 8.5

   Pure RNA has an A_{260}/A_{280} ratio of 1.9 – 2.1 in Tris-Cl, pH 7.5

2. DNA molar conversions

   1 ug of 1000 bp DNA = 1.52 pmole

   1 ug of pBR322 DNA = 0.36 pmole DNA

   1 pmole of pBR322 DNA = 2.8 ug

3. Protein molar conversions

   100 pmole of 100,000 MW protein = 10 ug

   100 pmole of 50,000 MW protein = 5 ug

   100 pmole of 10,000 MW protein = 1 ug

4. Protein/DNA conversions

   1kb DNA = 333 amino acids coding capacity = 10,000 MW protein

   = 6.5×10^5 Daltons of duplex DNA (sodium salt) and 3.3×10^5 Daltons SS DNA (sodium salt) = 3.4×10^5 Daltons of SS RNA (sodium salt)

   1 Dalton = the atomic mass of one proton or electron

   Average MW of a DNA base = 324.5 Daltons, 1bp = 649 Daltons

   Units 1g = 10^3mg = 10^6ug = 10^9ng = 10^{12}pg

   Human DNA = 6 pg/cell, 1pg = 6.4×10^{11} Daltons, Plant DNA = 0.5 – 200 pg/cell

5. Half life of important radioisotopes used

   Tritium = 12.43 years

   Carbon-14 = 5730 years

Sulphur-35 = 87.4 days

Phosphorous-32 = 14.3 days

6. On Tris-acetate agarose gels relative positions of different DNA forms

The exact distance between bands is influenced by percentage of agarose, time of electrophoresis, concentration of Ethidium bromide, degree of supercoiling and the size and complexity of the DNA.

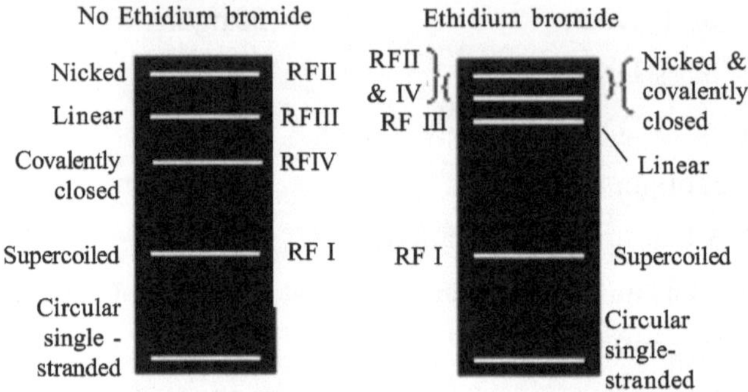

v) Codon chart and amino acid symbols and genetic code

Second base

First base	U		C		A		G		Third base
U	UUU	Phe	UCU	Ser	UAU	Tyr	UGU	Cys	U
	UUC		UCC		UAC		UGC		C
	UUA	Leu	UCA		UAA	Ocr	UGA	Opl	A
	UUG		UCG		UAG	Amb	UGG	Trp	G
C	CUU	Leu	CCU	Pro	CAU	His	CGU	Arg	U
	CUC		CCC		CAC		CGC		C
	CUA		CCA		CAA	Gln	CGA		A
	CUG		CCG		CAG		CGG		G
A	AUU	Ile	ACU	Thr	AAU	Asn	AGU	Ser	U
	AUC		ACC		AAC		AGC		C
	AUA		ACA		AAA	Lys	AGA	Arg	A
	AUG	Met	ACG		AAG		AGG		G
G	GUU	Val	GCU	Ala	GAU	Asp	GGU	Gly	U
	GUC		GCC		GAC		GGC		C
	GUA		GCA		GAA	Glu	CGA		A
	GUG		GCG		GAG		GGG		G

Termination Signals UAA (Ochre), UAG (Amber), UGA (Opal)

vi) Chemicals, kits and their sources

Chemical/kits	Sources
Vitamins; Hormonoes; Antibiotics and other molecular Biology grade chemicals	Sigma Chemical Company, St Louis, USA
Ethanol	Bengal Chemicals, Kolkata
Chemicals for preparation of MS medium and nutrient solutions	Hi Media Laboratoris, Mumbai
Restriction enzymes	MBI, Fermentas, Lithuania, USA
Taq. DNA polymerase and dNTPs.T4 DNA ligase,	Bangalore Genei, Bangalore
λP-GEM-T Easy vector kit DNA/*Hind*III and λ DNA *Hind*III + *EcoRI* markers, 1 kb ladder	MBI Fermentas, Lithuania, USA
P-GEM-T Easy vector kit	Promega Corporation, Wisconsin, USA
Plasmid Midiprep kit, PCR purification kit, Gel extraction kit	Qiagen Inc. Valencia, USA
Primers	Labmate Asia and Genuine Chemical Coproration, Delhi
Plasticwares	Tarsons, Kolkata and Axygen, New Delhi
Glasswares	Schott, Germany and Vensil, Bangalore
Pfu DNA polymerase, Reverse transcriptase	MBI, Fermentas Lithuania, USA